EARTH SCIENCES IN THE 21ST CENTURY

ICE-CAVES OF FRANCE AND SWITZERLAND

EARTH SCIENCES IN THE 21ST CENTURY

Additional books and e-books in this series can be found on Nova's website under the Series tab.

EARTH SCIENCES IN THE 21ST CENTURY

ICE-CAVES OF FRANCE AND SWITZERLAND

GEORGE FORREST BROWNE

snova
New York

NOTICE TO THE READER

The Publisher has taken reasonable care in the preparation of this book, but makes no expressed or implied warranty of any kind and assumes no responsibility for any errors or omissions. No liability is assumed for incidental or consequential damages in connection with or arising out of information contained in this book. The Publisher shall not be liable for any special, consequential, or exemplary damages resulting, in whole or in part, from the readers' use of, or reliance upon, this material. Any parts of this book based on government reports are so indicated and copyright is claimed for those parts to the extent applicable to compilations of such works.

Independent verification should be sought for any data, advice or recommendations contained in this book. In addition, no responsibility is assumed by the Publisher for any injury and/or damage to persons or property arising from any methods, products, instructions, ideas or otherwise contained in this publication.

This publication is designed to provide accurate and authoritative information with regard to the subject matter covered herein. It is sold with the clear understanding that the Publisher is not engaged in rendering legal or any other professional services. If legal or any other expert assistance is required, the services of a competent person should be sought. FROM A DECLARATION OF PARTICIPANTS JOINTLY ADOPTED BY A COMMITTEE OF THE AMERICAN BAR ASSOCIATION AND A COMMITTEE OF PUBLISHERS.

Additional color graphics may be available in the e-book version of this book.

Library of Congress Cataloging-in-Publication Data

ISBN: 978-1-53617-708-4

Published by Nova Science Publishers, Inc. † New York

CONTENTS

List of Illustrations

PREFACE[*]

The existence of natural ice-caves at depths varying from 50 to 200 feet below the surface of the earth, unconnected with glaciers or snow mountains, and in latitudes and at altitudes where ice could not under ordinary circumstances be supposed to exist, has attracted some attention on the Continent; but little or nothing seems to be practically known in England on the subject. These caves are so singular, and many of them so well repay inspection, that a description of the twelve which I have visited can scarcely, as it seems to me, be considered an uncalled-for addition to the numerous books of travel which are constantly appearing. In order to prevent my narrative from being a mere dry record of natural phenomena, I have interspersed it with such incidents of travel as may be interesting in themselves or useful to those who are inclined to follow my steps. I have also given, from various sources, accounts of similar caves in different parts of the world.

A pamphlet on *Glacières Naturelles* by M. Thury, of Geneva, of the existence of which I was not aware when I commenced my explorations, has been of great service to me. M. Thury had only visited three glacières when he published his pamphlet in 1861, but the observations he records

---

[*] This is an edited, augmented and reformatted version of "Ice-Caves of France and Switzerland" by Rev. G. F. Browne, MA, originally published by Longmans, Green and Co., dated 1865.

are very valuable. He had attempted to visit a fourth, when, unfortunately, the want of a ladder of sufficient length stopped him.

I was allowed to read Papers before the British Association at Bath (1864), in the Chemical Section, on the prismatic formation of the ice in these caves, and in the Geological Section, on their general character and the possible causes of their existence.

It is necessary to say, with regard to the sections given in this book, that, while the proportions of the masses of ice are in accordance with measurements taken on the spot, the interior height of many of the caves, and the curves of the roof and sides, are put in with a free hand, some of them from memory. And of the measurements, too, it is only fair to say that they were taken for the most part under very unfavourable circumstances, in dark caves lighted by one, or sometimes by two candles, with a temperature varying from slightly above to slightly below the freezing-point, and with no surer foot-hold than that afforded by slippery slopes of ice and chaotic blocks of stone. In all cases, errors are due to want of skill, not of honesty; and I hope that they do not generally lie on the side of exaggeration.

Cambridge: *June 1865.*

Chapter 1

THE GLACIÈRE OF LA GENOLLIÈRE, IN THE JURA

In the summer of 1861, I found myself, with some members of my family, in a small rustic *pension* in the village of Arzier, one of the highest villages of the pleasant slope by which the Jura passes down to the Lake of Geneva. The son of the house was an intelligent man, with a good knowledge of the natural curiosities which abound in that remarkable range of hills, and under his guidance we saw many strange things. More than once, he spoke of the existence of a *glacière* at no great distance, and talked of taking us to see it; but we were skeptical on the subject, imagining that *glacière* was his patois for *glacier*, and knowing that anything of the glacier kind was out of the question. At last, however, on a hot day in August, we set off with him, armed, at his request, with candles; and, after two or three hours of pine forests, and grass glades, and imaginary paths up rocky ranges of hill towards the summits of the Jura, we came to a deep natural pit, down the side of which we scrambled. At the bottom, after penetrating a few yards into a chasm in the rock, we discovered a small low cave, perfectly dark, with a flooring of ice, and a pillar of the same material in the form of a headless woman, one of whose shoulders we eventually carried off, to regale our parched friends at Arzier. We lighted up the cave with candles, and sat crouched on the ice drinking our wine, finding water, which served the double purpose of icing and diluting the wine, in small basins in the floor of ice, formed apparently by drops falling from the roof of the cave.

A few days after, our guide and companion took us to an ice-cavern on a larger scale, which, we were told, supplies Geneva with ice when the ordinary stores of that town fail; and the next year my sisters went to yet another, where, however, they did not reach the ice, as the ladder necessary for the final drop was not forthcoming.

In the course of the last year or two, I have mentioned these glacières now and then in England, and no one has seemed to know anything about them; so I determined, in the spring of 1864, to spend a part of the summer in examining the three we had already seen or heard of, and discovering, if possible, the existence of similar caves.

The first that came under my notice was the Glacière of La Genollière; and, though it is smaller and less interesting than most of those which I afterwards visited, many of its general features are merely reproduced on a larger scale in them. I shall therefore commence with this cave, and proceed with the account of my explorations in their natural order. It is probable that some of the earlier details may seem to be somewhat tedious, but they are necessary for a proper understanding of the subject.

La Genollière is the *montagne*, or mountain pasturage and wood, belonging to the village of Genollier, an ancient priory of the monks of S. Claude.[1] The cave itself lies at no great distance from Arzier--a village which may be seen in profile from the Grand Quai of Geneva, ambitiously climbing towards the summit of the last slope of the Jura. To reach the cave from Geneva, it would be necessary to take train or steamer to Nyon, whence an early omnibus runs to S. Cergues, if crawling up the serpentine

[1] In this neighbourhood, the *montagne* of any *commune* is represented by the feminine form of the name of the village: thus, *L'Arzière* is the *montagne* of Arzier, and *La Bassine* of Bassin. This has a curious effect in the case of some villages—such, for instance, as S. Georges—one of the landmarks of the district between the lakes of Joux and Geneva being the *Châlet de la S. Georges*, a grammatical anomaly which puzzles a stranger descending the southernmost slope of the Jura from the Asile de Marchairuz. This law of formation is not universal; for the *montagnes* of Rolle and S. Livres are called the *Prè de Rolle* and the *Prè de S. Livres*, while the *Fruitière de Nyon* is the rich upland possession of the town of that name.

road can be called running; and from S. Cergues a guide must be taken across the Fruitière de Nyon, if anyone can be found who knows the way. From Arzier, however, which is nine miles up from Nyon, it was not necessary to take the S. Cergues route; and we went straight through the woods, past the site of an old convent and its drained fish-pond, and up the various rocky ridges of hill, with no guide beyond the recollection of the previous visits two and three years before, and a sort of idea that we must go north-west. As it was not yet July, the cows had not made their summer move to the higher châlets, and we found the mountains uninhabited and still.

The point to be made for is the upper Châlet of La Genollière, called by some of the people *La Baronne*,[2] though the district map puts La Baronne at some distance from the site of the glacière. We had some difficulty in finding the châlet, and were obliged to spread out now and then, that each might hunt a specified portion of the wood or glade for signs to guide our further advance, enjoying meanwhile the lilies of the mountain and lilies of the valley, and fixing upon curious trees and plants as landmarks for our return. In crossing the last grass, we found the earliest vanilla orchis (*Orchis nigra*) of the year, and came upon beds of moonwort (*Botrychium Lunaria*) of so unusual a size that our progress ceased till such time as the finest specimens were secured.

Sometime before reaching this point, we caught a glimpse of a dark speck on the highest summit in sight, which recalled pleasantly a night we had spent there three years before for the purpose of seeing the sun rise.[3] My sisters had revisited the Châlet des Chèvres, which this dark speck represented, in 1862, and found that the small chamber in which we had slept on planks and logs had become a more total ruin than before, in the course of the winter, so that it is now utterly untenable.

---

[2] Probably a relic of the time when the earlier Barons of Coppet possessed this district. The families of Grandson, Lesdiguières, and Dohna successively held the barony; and in later times the title *de Coppet* hid a name more widely known, for on the Châlet of *Les Biolles*, some distance to the east of La Baronne, the name of *Auguste de Staël de Holstein de Coppet* is carved, after the fashion of Swiss châlets. This was Madame de Staël's son, who built Biolles in 1817; it was afterwards sold to the commune of Nyon, and finally purchased by Arzier two or three years ago.

[3] 'Cornhill Magazine,' June 1863, 'How we slept at the Châlet des Chèvres.'

From Arzier to the Châlet of La Genollière, would be about two hours, for a man walking and mounting quickly, and never losing the way; and the glacière lies a few minutes farther to the north-west, at an elevation of about 2,800 feet above the lake, or 4,000 feet above the sea.[4] A rough mountain road, leading over an undulating expanse of grass, passes narrowly between two small clumps of trees, each surrounded by a low circular wall, the longer diameter of the enclosure on the south side of the road being 60 feet. In this enclosure is a natural pit, of which the north side is a sheer rock, of the ordinary limestone of the Jura, with a chasm almost from the top; while the south side is less steep, and affords the means of scrambling down to the bottom, where a cave is found at the foot of the chasm, passing under the road. The floor of this small but comparatively lofty cave is 52 feet below the surface of the earth, and slopes away rapidly to the west, where, by the help of candles, the rock which forms the wall is seen to stop short of the floor, leaving an entrance 2 or 3 feet high to an inner cave--the glacière. The roof of this inner cave rises slightly, and its floor falls, so that there is a height of about 6 feet inside, excepting where a large open fissure in the roof passes high up towards the world above. At one end, neither the roof nor the floor slopes much, and in this part of the cave the height is less than 3 feet.

It would be very imprudent to go straight into an ice-cave after a long walk on a hot summer's day, so we prepared to dine under the shade of the trees at the edge of the pit, and I went down into the cave for a few moments to get a piece of ice for our wine. My first impression was that the glacière was entirely destroyed, for the outer cave was a mere chaos of rock and stones; but, on further investigation, it turned out that the ruin had not reached the inner cave. In our previous visit we had noticed a natural basin of some size and depth among the trees on the north side of the road, and we now found that the chaos was the result of a recent falling-in of this basin; so that from the bottom of the first cave, standing as it were under the road, we could see daylight through the newly-formed hole.

---

[4] This is only a guess, made from a comparison with the ascertained heights of neighbouring points.

The total length of the floor of the inner cave, which lies north-east and south-west, is 51 feet; and of this floor a length of about 37 feet was more or less covered with ice, the greatest breadth of the ice being within an inch or two of 11 feet. Excepting in the part of the cave already mentioned as being less than 3 feet high, we found the floor not nearly so dry, nor so completely covered with ice, as when we first saw the glacière, three years before, in the middle of an exceptionally hot August. Under the low roof all was very dry, though even there the ice had not an average thickness of more than 8 inches. It may be as well to say, once for all, that the ice in these caves is never found in a sheet on a pool of water; it is always solid, forming the floor of the cave, filling up the interstices of the loose stones, and rising above them, in this case with a surface perfectly level.

Figure 1. Ice-Columns in the Glacière Of La Genollière.

We found four principal columns of ice, three of which, in the loftiest part of the cave, are represented in the accompanying engraving: I call them three, and not two, because the two which unite in a common base proceeded from different fissures. The line of light at the foot of the rock-wall is the only entrance to the glacière. The lowest column was 11-2/3 feet high and 1-2/3 feet broad, not more than 6 inches thick in the middle, half-way up, and flattened symmetrically so as to be comparatively sharp at the edges, like a huge double-edged sword. It stood clear of the rock through its whole height, but scarcely left room between itself and the wall of the cave for a candle to be passed up and down. The other two columns shown in the engraving poured out of fissures in the rock, streaming down as cascades, the one being 13-1/2 and the other 15 feet high; and when we tied a candle to the end of an alpenstock, and passed it into thc fissures, we found that the bend of the fissures prevented our seeing the termination of the ice. An intermittent disturbance of the air in these fissures made the flame flicker at intervals, though generally the candle burned steadily in them, and we could detect no current in the cave. The fourth column was in the low part of the cave, and we were obliged to grovel on the ice to get its dimensions: it was 3-1/4 feet broad and 4-1/3 feet high, the roof of the cave being only 2-3/4 feet high; and it poured out of the vertical fissure like a smooth round fall of water, adhering lightly to the rock at its upper end like a fungus, and growing out suddenly in its full size. This column was dry, whereas on the others there were abundant symptoms of moisture, as if small quantities of water were trickling down them from their fissures, though the fissures themselves appeared to be perfectly dry.

In one of the fissures there was a patch of what is known as sweating-stone,[5] with globules of water oozing out, and standing roundly upon it: the globules were not frozen. This stone was exceedingly hard, and defied all our efforts to break off a specimen, but at last we got two small pieces, hard and heavy, and wrapped them in paper; ten weeks after, we found them of course quite dry, and broke them easily, small as they were, with our fingers. The fissure from which the shortest of the four columns came was full of gnats, as were also several crevices in the walls of the cave, especially in the lowest part; and we found a number of large red-brown

[5] The patois of Vaud has a prettier name for this kind of stone—*le sex* (or *scex) qui plliau*, the weeping-stone.

flies,[6] nearly an inch long, running rapidly on the ice and stones, after the fashion of the flies with which trout love best to be taken. The central parts of the cave, where the roof is high, were in a state provincially known as 'sloppy,' and drops of water fell now and then from above, either splashing on wet stones, or hollowing out basins in the remaining ice, or, sometimes, shrewdly detecting the most sensitive spot in the back of the human neck. We placed one of Casella's thermometers on a piece of wood on one of the wet stones, clear of the ice, and it soon fell to 34°. Probably the temperature had been somewhat raised by the continued presence of three human beings and two lighted candles in the small cavern; and, at any rate, the cold of two degrees above freezing was something very real on a hot summer's day, and told considerably upon my sisters, so that we were compelled to beat a retreat,--not quite in time, for one of our party could not effect a thaw, even by stamping about violently in the full afternoon sun.

While we were in the cave, we noticed that the surfaces of the columns were covered by very irregular lines, marked somewhat deeply in the ice, and dividing the surface into areas of all shapes, a sort of network, with meshes of many different shapes and sizes. These areas were smaller towards the edges of the columns; the lines containing them were not, as a rule, straight lines, and almost baffled our efforts to count them, but, to the best of my belief, there were meshes with three, four, and up to eight sides. The column which stood clear of the rock was composed of very limpid ice, without admixture of air; but the cascades were interpenetrated by veins of looser white ice, and, where the white ice came, the surface lines seemed to disappear. As we sat on the grass outside, arranging our properties for departure, my attention was arrested by the columnar appearance of the fractured edge of the block of ice which we had used at luncheon. It was about 5 inches thick, and had formed part of a stalagmite whose horizontal section, like that of the free column, would be an ellipse of considerable eccentricity; and, on examination, it turned out that the surface areas, which varied in size from a large thumb-nail to something

---

[6] I brought one of these to England, and am told that it is the *Stenophylax hieroglyphicus* of Stephens, or something very like that fly.

very small, were the ends of prisms reaching through to the other side of the piece of ice, at any rate in the thinner parts, and presenting there similar faces. Not only so, but the prisms could be detached with great ease, by using no instrument more violent than the fingers; while the point of a thin knife entered freely at any of the surface lines, and split the ice neatly down the sides of the prisms. When one or two of the sides of a prism were exposed, at the edge of the piece of ice, the prism could be pushed out entire, like a knot from the edge of a piece of wood. In some cases there seemed to be capillary fissures coincident with the lines where several sides of prisms met. Considering the shape of the whole column, it is clear that the two ends of each prism could not be parallel; neither was one of the ends perfectly symmetrical with the other, and I do not think that the prisms were of the nature of truncated pyramids. On descending again, I found that the columns were without exception formed of this prismatic ice, either in whole, as in the clear column, or in part, as where limpid prisms existed among the white ice which ran in veins down the cascades. In the free vertical column the prisms seemed to be deposited horizontally, and in the thicker parts they did not pass clear through. We carried a large piece of ice down to Arzier in a botanical tin, and on our arrival there we found that all traces of external lines had disappeared.

This visit to the glacière was on Saturday, and on the following Monday I determined to go up alone, to take a registering thermometer, and leave it in the cave for the night; which, of course, would entail a third visit on the next day. Monday brought a steady penetrating rain, of that peculiar character which six Scotch springs had taught me to describe as 'just a bit must;' while in the higher regions the fog was so hopeless, that a sudden lift of the mist revealed the unpleasant fact that considerable progress had been made in a westerly direction, the true line being north-west. Instead of the rocks of La Genollière, the foreground presented was the base of the Dôle, and the chasm which affords a passage from the well-known fortress of Les Rousses into Vaud. There was nothing for it but to turn in the right direction, or attempt to do so, and force a way through the wet woods till something should turn up. This something took the form of a châlet; but no amount of hammering and shouting produced any response, and it was only after a forcible entrance, and a prolonged course of interior shouting, that a man was at length drawn. He said that he had been asleep--and why he put it in a past tense is still a mystery--and could give no idea of the direction of the châlet on La Genollière, beyond a vague suggestion that it was somewhere in the mist; a suggestion by no means improbable, seeing that the mist was ubiquitous. One piece of information he was able to give, and it was consoling: I was now, it seemed, on the Fruitière de Nyon, and therefore the desired châlet could not be far off, if only a guide could be found. On the whole, he thought that a guide could not be found; but there were men in the châlet, and I might go up the ladder with him and see what could be done. He led to a chamber with a window of one small pane, dating apparently from the first invention of glass, and never cleaned since. An invisible corner of the room was appealed to; but the voice which resided there, and seemed like everything else to be asleep, pleaded dreamily a total ignorance of the whereabouts of the châlet in question. Just as, by dint of steady staring through the darkness, an indistinct form of a mattress, with a human being reclining thereon, began to be visible, another dark corner announced that this new speaker had heard of a *p'tit sentier* leading to the châlet, but knew neither direction nor distance. Here the space between the two corners put in a word; and, as the darkness was now becoming natural, seven or eight mattresses appeared, ranged round the room, some holding one, some two men, most of whom were sitting up on end with old caps on, displaying every variety of squalor. The voice which had spoken last declared that the distance was three-quarters of an hour, and that if the day were clear there would be no difficulty in reaching the châlet; as it was, the man would be very glad to try.

A change of cap was the only dressing necessary for the volunteer, and we faced the fog and rain, which elicited from him such a disgraceful amount of swearing, that it was on all accounts well when the rain ceased for a few minutes, the mists rolled off, and the clouds lifted sufficiently to betray the surface of the Lake of Geneva, luxuriating in the clear warmth of an early summer's day, and making us shiver by the painful contrast which our own altitude presented. The deep blue of the lake brought to mind the story of the shepherd of Gessenay (Saanen), of whom it is told that when he was passing the hills with some friends for a first visit to Vevey, and came in sight of the lake, which he had never seen before, he turned and hurried home incontinent, declaring that he would not enter a country where the good God had made the blue sky to fall and fill the valleys.

In this bright interval we came upon a magnificent fox, and the peasant's impulse was, 'Oh, for a good gun!' an exclamation which would have sounded horrible to English ears, if I had not been previously broken in to it by an invitation from a Scotch gamekeeper to a fox-hunt, when he promised an excellent gun, and a *stance* which the foxes were sure to pass.

The rain now came on again, and the guide thought he had had plenty of it, and must return for the afternoon milking; and just then, as good luck would have it, we stumbled upon an immense clump of nettles which had been one of our landmarks two days before, so that he was no longer necessary, and we said affectionate adieux.

The glacière was in a state of ruin. Only the right-hand column, not speaking heraldically, was standing, the others lying in blocks frozen hard together on the ground. The column which still stood was much shrunken, and seemed too small for its fissure, the sides of which it scarcely touched. The wind blew down the entrance slope so determinedly, that a candle found it difficult to live at the bottom of the first cave; and a portion of the current blew into the glacière, and in its sweep exactly struck the fallen columns, the edges of which were already rounded by thaw. Much of this must be attributed to the recent opening of the second shaft (p. 5), which admits a thorough draught through the first cave, and so exposes the glacière to currents of warmer air; and I should expect to find that in future the ice will disappear from that part of the cave every summer,[7] whereas in 1861 we found it thick and dry (excepting a few small basins containing water) and evidently permanent, in the middle of a very hot August. The low part of the cave was so completely protected from the current, that the candle burned there quite steadily for an hour and a half: still, like the

[7] Since writing this, I have been told that some English officers who visited the cave in the August of 1864 found no ice in any part.

others, the column at that end of the glacière was broken down, and it therefore became necessary to attribute its fall to some other agency than the current of external air. There had been a very large amount of rain, and the surface of the rock in the fissures was evidently wet; so I have no doubt that the filtering through of the warm rain-water had thawed the upper supports of the ice-cascades, and then, owing to their slightly inclined position, the pedestal had not provided sufficient support, and so they had fallen. One of them, perhaps, had brought down in its fall the free column, which had stood two days before on its own base, without any support from the rock. Very probably, too--indeed, almost certainly,--the fall of the large mass of rock, which once formed the bottom of the basin on the north side of the road, has affected the old-established fissures, by which rain-water has been accustomed to penetrate in small quantities to the glacière, so that now a much larger amount is admitted. On this account, there will probably be a great diminution of the ice in the course of future summers, though the amount formed each winter may be greater than it has hitherto been. Constant examination of other columns and fissures has convinced me, that, before the end of autumn, the majority of the glacières will have lost all the columns which depend upon the roof for a part of their support, or spring from fissures in the wall; whereas those which are true stalagmites, and are self-supporting, will have a much better chance of remaining through the warm season, and lasting till the winter, and so increasing in size from year to year. Free stalagmites, however, which are formed under fissures capable of pouring down a large amount of water on the occasion of a great flood of rain, must succumb in time, though not so soon as the supported columns.

A curious appearance was presented by a small free stalagmite in the retired part of the cave. The surface of the stalagmite was wet, from the drops proceeding from a fissure above, and was lightly covered in many parts with a calcareous deposit, brought down from the fissures in the roof by the water filtering through. The stalagmite was of the double-edged-sword shape, and the limestone deposit collected chiefly at one of its edges, the edge nearer to that part of the cave where thaw prevailed; so that the real edge was a ridge of deposit beyond the edge of the ice.[8] Patches of limestone paste lay on many parts of the ice-floor.

In the loftier part of the cave, water dropped from the roof to so large an extent, that ninety-six drops of water in a minute splashed on to a small stone immediately under the main fissure. This stone was in the centre of a considerable area of the floor which was clear of ice; and it struck me that if the columns were formed by the freezing of water dropping from the roof, there ought to have been at some time a large column under this, the most plentiful source of water in the cave. Accordingly, I found that the edge of the ice round this clear area was much thicker than the rest of the ice of the floor, and was evidently the remains of the swelling pedestal of a column which had been about 12 feet in circumference. This departed column may account for a fact which I discovered in another glacière, and found to be of very common occurrence, viz., that in large stalagmites there is a considerable internal cavity, extending some feet up from the ground, and affording room even for a man to walk about inside the column. When the melted snows of spring send down to the cave, through the fissures of the rock, an abundance of water at a very low temperature, and the cave itself is stored with the winter's cold, these thicker rings of ice catch first the descending water, and so a circular wall, naturally conical, is formed round the area of stones; the remaining water either running off through the interstices, or forming a floor of ice of less thickness, which yields to the next summer's drops. In the course of time, this conical wall rises, narrowing always, till a dome-like roof is at length formed, and thenceforth the column is solid. Of course, the interior cannot be wholly free from ice; and it will be seen from the account of one of these cavities, which I explored in the Schafloch, that they are decked with ice precisely as might be expected.[9] Another possible explanation of this curious and beautiful phenomenon will be given hereafter.[10]

The temperature was half a degree lower than when there were three of us in the cave two days before. I deposited one of Casella's registering

---

[8] See also p. 231.
[9] P. 145.
[10] P. 301.

thermometers, on wood, on a stone in that part of the floor which was free from ice, though there was ice all round it at some little distance. The thermometer was well above the surface of the ice, and was protected from chance drops of water from the roof.

The next morning I started early from Arzier, having an afternoon journey in prospect to the neighbourhood of another glacière, and was accompanied by Captain Douglas Smith, of the 4th Regiment. On our way to La Genollière, we came across the man who had served as guide the day before, and a short conversation respecting the glacière ensued. He had only seen it once, many years before, and he held stoutly to the usual belief of the peasantry, that the ice is formed in summer, and melts in winter; a belief which everything I had then seen contradicted. His last words as we parted were, '*Plus il fait chaud, plus ça gèle*;' and, paradoxical as it may appear, I believe that some truth was concealed in what he said, though not as he meant it. Considering that his ideas were confined to his cattle and their requirements, and that water is often very difficult to find in that part of the Jura, a *hot* summer would probably mean with him a *dry* summer, that is, a summer which does not send down much water to thaw the columns in the cave. Extra heat in the air outside, at any season, does not, as experience of these caves proves abundantly, produce very considerable disturbance of their low temperature, and so summer water is a much worse enemy than extra summer heat; and if the caves could be protected from water in the hot season, the columns in them would know how to resist the possible--but very small--increase of temperature due to the excess of heat of one summer above another. And since the eye is most struck by the appearance of the stalagmites and ice-cascades, it may well be that the peasants have seen these standing at the end of an unusually hot and dry summer, and have thence concluded that hot summers are the best time for the formation of ice. Of course, at the beginning of the winter after a hot summer, there will be on these terms a larger nucleus of ice; and so it will become true that the hotter the year, the more ice there will be, both during the summer itself and after the following winter.

The further process of the formation of ice will be this:--the colds of early winter will freeze all the water that may be in the glacières from the summer's thaw, in such caves as do not possess a drainage, and then the frost will have nothing to occupy itself upon but the ice already formed, for no water can descend from the frost-bound surface of the earth.[11] As soon as the snow begins to melt to so great a degree that the fissures are opened up once more, the extremely cold water resulting therefrom will descend through the limestone into a cave perfectly dry, and filled with an atmosphere many degrees below the freezing point, whose frost-power eagerly lays hold of every drop of water which does not make its escape in time by the drainage of the cave. Thus the spring months will be the great time of the formation of ice, and also of the raising of the temperature from some degrecs below freezing to the more temperate register at which I have generally found it, viz., rather above than below 32°. Professor Tyndall very properly likens the external atmosphere to a ratchet-wheel, from its property of allowing the passage of hot rays down to the surface of the earth, and resisting their return: it may equally be so described on other grounds, inasmuch as the cold and heavy atmosphere will sink in the winter into the pits which lead to glacières, and will refuse to be altogether displaced in summer by anything short of solar radiation.

We found the one column of the previous day still standing, though evidently in an unhappy state of decay. The sharpness of its edges was wholly gone, and it was withered and contorted; there were two cracks completely through it, dividing it into three pieces 4 or 5 feet long, which were clearly on the point of coming down. Externally, the day was fine and warm, and so we found the cave comparatively dry, only one drop falling in a minute on to the stone where ninety-six had fallen in the same time the day before. The thermometer registered 32° as the greatest cold of the night, and still stood at that point when we took it up.

---

[11] It is possible that the freezing of the surface may play a curious part in the phenomena of the spring season in such caves. Supposing the surface to be completely frost-bound, all atmospheric pressure will be removed from the upper surface of the water in the long fissures, and thus water may be held in suspension, in the centre of large masses of fissured rock, during the winter months. The first thorough thaw will have the same effect as the removal of the thumb from the upper orifice in the case of the hand-shower-bath; and the water thus rained down into the cave will have a temperature sufficiently high to destroy some portion of the cold stored up by the descent of the heavy atmosphere of winter, or at least to melt out the ice which may have blocked up the lower ends of the fissures.

We spent some little time in exploring the neighbourhood of the pits, in order to find, if possible, the outlet for the drainage, but the ground did not fall away sufficiently for any source from so low an origin to show itself. The search was suggested by what I remembered of the Glacière of S. Georges three years before, where the people believe that a small streamlet which issues from the bottom of a steep rock, some distance off, owes its existence to the glacière.

Chapter 2

THE GLACIÈRE OF S. GEORGES, IN THE JURA

The best way of reaching this glacière from Geneva would be to take the steamer to Rolle, or the train to one of the neighbouring stations, between Geneva and Lausanne, and thence pass up the slope of the Jura by the road which leads through Gimel. For the train, the Allaman station would be the most convenient, as an omnibus runs from Allaman to Aubonne, where the poste for Gimel may be caught. But from Arzier there is a short cut of less than two hours along the side of the hills, leaving that village by a deep gorge not unfitly named *L'Enfer*, and a dark wood which retains an odour of more savage bygone times in its name of the 'Bear's Wood,' as containing a cavern where an old bear was detected in the act of attempting to winter.[12]

The village of S. Georges has very respectable accommodation for a single traveller, *au Cavalier*. The common day-room will be found

---

[12] On our previous visit, in 1861, we passed from Arzier through Longirod and Marchissy, stopping to measure and admire the huge lime-tree in the churchyard of the latter village. Our Swiss companion on that occasion was anxious that we should carry home some ice from the cave; and as the communal law forbade the removal of the ice by strangers, he hunted up a cousin in Marchissy, and sent him with a *hotte* across country, while we went innocently by the ordinary route through S. Georges. The cousin, however, contrived to lose himself in the woods, and we never heard of him again.

untenable by most Englishmen, however largely they may delight in rough quarters; but there is a double-bedded room at the end of a bricked passage up-stairs, which serves well for bedroom and sitting-room in one. The chief drawback in this arrangement is, that the landlady inexorably removes all washing apparatus during the day, holding that a pitcher and basin are unseemly ornaments for a sitting-room. The deal table, of course, serves both for dressing and for feeding purposes, but it is fortunately so long that an end can be devoted to each; and on the whole it is possible to become considerably attached to the room, with its three airy windows, and the cool unceasing hum of a babbling fountain in the village-street below. The Auberge is a large building, with a clock-tower of considerable height, containing the clock of the commune: as soon as the candle is put out at night, it becomes painfully evident that a rectangular projection in one corner of the room is in connection with this tower, and in fact forms a part of the abode of the pendulum, which plods on with audible vigour, growing more and more audible as the hours pass on, and making a stealthy pervading noise, as if a couple of lazy ghosts were threshing phantom wheat. The clocks of Vaud, too, are in the habit of striking the hour twice, with a short interval; so that if anyone is not sure what the clock meant the first time, he has a second chance of counting the strokes. This is no doubt an admirable plan under ordinary circumstances, but it does certainly try the patience of a sleepless dyspeptic after a surfeit of café-au-lait and honey; and when he has counted carefully the first time, and is bristling with the consciousness that it is only midnight, it is aggravating in the extreme to have the long slow story told a second time within a few feet of his head.

The Cavalier had retained a guide overnight, Henri Renaud by name, and he appeared punctually at eight o'clock in the morning, got up in the short-tail coat of the country, and a large green umbrella with mighty ribs of whalebone. The weather was extremely unpleasant, a cold pitiless rain rendering all attempts at protection unavailing; but, fortunately, the glacière is only an hour and a quarter from the village. The path is tolerably steep, leading across the *petit Pré de Rolle*, and through woods of beech and fir, till the summit of one of the minor ridges of the Jura is reached, whence a short descent leads to the mouth of the glacière, something more than 4,000 feet above the sea. The ground here slopes down towards the north; and on the slope, among fir-trees, an irregular circular basin is seen, some seven or eight yards across[13] and perhaps two yards deep, at the bottom of which are two holes. One of these holes is open, and as the guide and I--for my sisters remained at Arzier--stood on the neck of ground between the holes, we could see the snow lying at the bottom of the cave; the other is covered with trunks of trees, laid over the mouth to prevent the rays of the sun from striking down on to the ice. This protection has become necessary in consequence of an incautious felling of wood in the immediate neighbourhood of the mouth, which has exposed the ice to the assaults of the weather. The commune has let the glacière for a term of nine years, receiving six or seven hundred francs in all; and the *fermier* extracts the ice, and sells it in Geneva and Lausanne. In hot summers, the supplies of the artificial ice-houses fail; and then the hotel-keepers have recourse to the stores laid up for them by nature in the Glacières of S. Georges and S. Livres. Hence the importance of protecting the ice; the necessity for so doing arising in this case from the fact that the entrance to the cave is by a hole in the roof, which exposes the ice to direct radiation, unlike all other glacières, excepting perhaps the *Cueva del Hielo* on the Peak of Teneriffe.[14]

Autumn appears to be the usual time for cutting the ice, when it is carried from the cave on men's backs as far as the commencement of the rough mountain-road, and is there packed on chars, and so conveyed to the nearest railway station. Renaud had worked in the cave for two years, and asserted that they did not choose the night for carrying the ice down to the station, and did not even care to choose a cool day. He believed that, in the autumn of 1863, they loaded two chars a day for fifteen days, and each char took from 40 to 50 quintaux; the quintal containing 50 kilos, or 100 livres.[15] In Professor Pictet's time (1822) this glacière supplied the Hospital of Geneva, whose income depended in part on its privilege of *revente* of all

---

13 The size of this basin is exaggerated in the engraving on page 24, owing to the roughness of the original sketch.

14 See p. 253.

15 For further details on this point see pages 54 and 83.

ice sold in the town, with 25 quintaux every other day during the summer. In my anxiety to learn the exact amount of ice now supplied by the glacière, I determined to find out the *fermier*; but Renaud could tell nothing of him beyond the fact that he lived in Geneva, which some promiscuous person supplemented by the information that his name was Boucqueville, and that he had something to do with comestibles. On entering upon a hunt for M. Boucqueville a fortnight later, it turned out that no one had heard of such a person, and the Directory professed equal ignorance; but, under the head of 'Comestibles,' there appeared a Gignoux-Bocquet, No. 34, Marché. Thirty-four, Marché, said, yes--M. Bocquet--it was quite true: nevertheless, it was clear that monsieur meant Sebastian aîné, on the Molard. The Molard knew only a younger Sebastian, but suggested that the right man was probably M. Gignoux-Chavaz, over the way; and when it was objected that Gignoux-Bocquet, and not Gignoux-Chavaz, was the name, the Molard replied that it made no matter,--Chavaz or Bocquet, it was all the same. When M. Gignoux-Chavaz was found, he said that he certainly was a man who had something to do with a glacière, but, instead of farming the Glacière of S. Georges, he had only bought a considerable quantity of ice two years ago from the Glacière of S. Livres, and he did not believe that the *fermier* of S. Georges lived in Geneva. Part of the confusion was due to the custom of placing a wife's maiden name after her husband's name: thus Gignoux-Chavaz implies that a male Gignoux has married a female Chavaz; and when a Swiss marries an English lady with a very English name, the result in the Continental mouth is sufficiently curious.

On arriving at the entrance to the glacière, the end of a suggestive ladder is seen under the protecting trunks; and after one or two steps have been taken down the ladder, the effect of the cave below is extremely remarkable, the main features being a long wall covered thickly with white ice in sheets, a solid floor of darker-coloured ice, and a high pyramid of snow reaching up towards the uncovered hole already spoken of. The atmosphere of the cave is damp, and this causes the ladders to fall speedily to decay, so that they are by no means to be trusted: indeed, an early round gave way under one of my sisters, when they visited the cave with me in

1861, and suggested a clear fall of 60 feet on to a cascade of ice.[16] There are three ladders, one below the other, and a hasty measurement gave their lengths as 20, 16, and 28 feet. The rock-roof is only a few feet thick in the neighbourhood of the hole of entrance.

Figure 2. Entrance to the Glacière of S. Georges.

[16] These ladders have at best but little stability, as they consist of two uprights, careless about the coincidence of the holes, with bars poked loosely through and left to fall out or stay in as they choose, the former being the prevailing choice. One of the ladders happened to be firmer than the generality of its kind; but, unfortunately, its legs were of unequal lengths, and so it turned round with one of my sisters, leaving her clinging like a cat to the underside. When the bars are sufficiently loose, a difference of a few inches in the lengths of the legs is not of so much importance.

The total length of the cave is 110 feet, lying NE. and SW., in the line of the main chain of the Jura. The lowest part of the floor is a sea of ice of unknown depth, 45 feet long by 15 broad; and Renaud tried my powers of belief by asserting that in 1834 the level of this floor was higher by half the height of the cave than now; a statement, however, which is fully borne out by Professor Pictet's measurements in 1822, when the depth of the glacière was less than 30 feet. Indeed, the floor had sunk considerably since my previous visit, when it was all at the same level down to the further end of the cave; whereas now, as will be seen in the section, there was a platform of stones resting on ice at that end. There are two large fissures passing into the rock, one only of which can be represented in the section, and these were full of white ice, not owing its whiteness apparently to the admixture of air in bubbles, but firm and compact, and very hard, almost like porcelain. Small stalactites hung from round fissures in the roof, formed of the same sort of ice, and broken off short, much as the end of a leaden pipe is sometimes seen to project from a wall. With this exception, there was no ice hanging from the roof, though there were abundant signs of very fine columns which had already yielded to the advancing warmth: one of these still remained, in the form of broken blocks of ice, in the neighbourhood of the open hole in the roof, immediately below which hole the stones of the floor were completely bare, and the thermometer stood at 50°. At the far end of the cave, the thermometer gave something less than 32°; a difference so remarkable, at the same horizontal level, that I am inclined to doubt the accuracy of the figures, though they were registered on the spot with due care. The uncovered hole, it must be remembered, is so large, and so completely open, that the rain falls freely on to the stones on the floor below.

By far the most striking part of this glacière is the north-west wall, which is covered with a sheet of ice 70 feet long, and 22 feet high at the highest part: in the neighbourhood of the ladders, this turns the corner of the cave, and passes up for about 9 feet under the second ladder. The general thickness of the sheet is from a foot to a foot and a half; and this is the chief source from which the *fermier* draws the ice, as it is much more easily quarried than the solid floor. Some of my friends went to the cave a few weeks after my visit, and found that the whole sheet had been pared off and carried away.

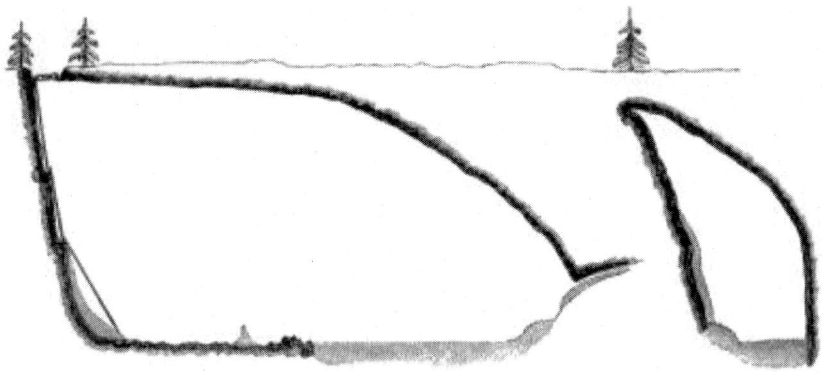

Figure 3. Vertical Sections of the Glacière of S. Georges.

On some parts of the wall the sheet was not completely continuous, being formed of broad and distinct cascades, connected by cross channels of ice, and uniting at their upper and lower ends, thus presenting many curious and ornamental groupings. On cutting through this ice, it was found not to lie closely on the rock, a small intermediate space being generally left, almost filled with minute limestone particles in a very wet state; and the whole cavern showed signs of more or less thaw.

It was natural to examine the structure of the ice in this glacière, after what we had observed on La Genollière. The same prismatic structure was universal in the sheet on the wall, and in the blocks which lay here and there on the floor and formed the sole remains of former columns. It was to be observed also in many parts of the ice-floor itself. The base of one large column still remained standing in its original position, and its upper end presented a tolerably accurate horizontal section of the column. The centre was composed of turbid ice, round which limpid prisms were horizontally arranged, diverging like the feathers of a fan; then came a ring of turbid ice, and then a second concentric ring of limpid prisms, diverging in the same manner as those which formed the inner ring. There were in all three or four of these concentric rings, the details showing a considerable amount of confusion and interference: the general law, however, was most evident, and has held in all the similar columns which I have since examined in other glacières. The rings were not accurately circular, but presented rather the appearance of having been formed round a roughly-fluted pillar on an elliptical base.

The examination of the ice on the wall gave some curious results. The horizontal arrangement of the prisms, which we had found to prevail in vertical columns, was here modified to suit the altered conditions of the case, and the axes of the prisms changed their inclination so as to be always perpendicular to the surface on which the ice lay, as far as could be determined by the eye. Thus, in following the many changes of inclination of the wall, the axes of the prisms stood at many different angles with the vertical, from a horizontal position where the wall chanced to be vertical, to a vertical position on the horizontal ledges of the rock. The extreme edges, too, of the ice, presented a very peculiar appearance. The general thickness, as has been said, varied from a foot to a foot and a half; and this diminished gradually along horizontal lines, till, at the edges of the sheet, where the ice ceased, it became of course nothing. The extreme edge was formed of globular or hemispherical beads of ice, like the freezing of a sweating-stone, lying so loosely on the rock that I could sweep them off in detail with one hand, and catch them with the other as they fell. Passing farther on towards the thicker parts of the ice, these beads stood up higher and higher, losing their roundness, and becoming compressed into prisms of all shapes, in very irregular imitation of the cellular tissue in plants, the axes of the prisms following the generally-observed law. There seems to be nothing in this phenomenon which cannot be accounted for by the supposition of gradual thaw of small amount being applied to a sheet of prismatic ice.

One fact was remarkable from its universal appearance. Wherever an incision was made in this sheet of ice, the prisms snapped off at the depth of an inch, and could be mowed down like corn by means of a stout knife. Although they broke naturally at this constant depth, and left a surface of limpid ice without any signs of external or internal division, still the laminae obtained by chiselling this lower surface carefully, broke up regularly into the shapes to be expected in sections of prisms cut at right angles to the axis. The roughness of my instruments made it impossible to discover how far this extended, and whether it ceased to be the case at any given depth in the ice.

The sea of ice on the floor was in a very wet state at the surface, being at a lower level than the stones on to which the rain from the open hole fell; and here the prismatic structure was not apparent to the eye, nor do I know whether it existed at all. In the Glacière of La Genollière I carried a large block of perfectly prismatic ice into the outer cave, where it was exposed to the free currents of air passing from the pit of entrance to the hole newly opened by the falling in of the ground; and, two days after, the external lines were scarcely perceptible, while on the occasion of our third visit I found that they had entirely disappeared, and the whole block was rapidly following their example. This disappearance of the surface-lines under the action of atmospheric thaw is probably the same thing as their absence when the flooring of ice is thinly covered with water. Wherever the flooring rose slightly towards the edges of the sea of ice, the usual structure appeared again.

There were no currents of air in the cave, the candles burning steadily through the whole time of our visit. Excepting for the purpose of detecting disturbance in the air, there is no need of candles, as the two holes in the roof supply sufficient light. Some account of the careful observations made here by M. Thury, at different seasons of the year, will be found in other parts of this book. We passed, on our return, by the source of water which springs from the foot of a rock at some distance from the glacière, and is supposed to form the outlet for the drainage of the cave; but it is difficult to understand how this can be the case, considering the form and character of the intervening ground.

The two ice-caves so far described are the least interesting of all that I have visited; but a peasant informed me, a day or two after, that if we had penetrated to the back of the pyramid of snow which lay half under the open hole, being the remains of the large collection which is formed there in the winter, we might have found a deep pit which is sometimes exposed by the melting of the snow. He had some idea that its depth was 30 feet a few years ago, and that its sides were solid ice. I shall have occasion to mention such pits in another glacière; if one does exist here, it has probably been quarried in the ice by the drops from the hole in the roof, and there might be some interest attached to an attempt to investigate it.[17]

We reached S. Georges again in a wretched state of wet and cold, and Renaud went off to bed, and imbibed abundant and super-abundant kirsch,--at least, when drawn thence the next morning, his manner left no doubt about either the fact or the abundance of the potations overnight. Warned by many experiences, I had gone no nearer to a specification of the bill of fare than a vague suggestion that *quelque chose* must be forthcoming, with an additional stipulation that this must be something more than mere onions and fat. The landlady's rendering of *quelque chose* was very agreeable, but, for the benefit of future diners *au Cavalier*, it is as well to say that those who do not like anisette had better make a private arrangement with their hostess, otherwise they will swallow with their soup an amount sufficient for many generations of the drag: they may also safely order savoury rice, with browned veal and wine-sauce, which is evidently a strong point with the Cavalier. All meals there are picturesque; for the omelette lay on the Castle of Grandson and a part of the Lake of Neufchâtel, while the butter reposed on the ruined Cathedral of Sion, and the honey distilled pleasantly from the comb on to the walls of Wufflens. No one should put any trust in the spoons, which are constructed apparently of pewter shavings in a chronic state of semi-fusion. On the evening of the second day, the landlady allowed a second knife at tea, as the knife-of-all-work had begun to knock up under the heavy strain upon its powers; but this supplementary instrument was of the ornamental kind, and, like other ornamental things, broke down at a crisis, which took the form of a piece of crust.

Lest this account should raise anyone's expectations too high, it is as well to add that they have no snuffers in S. Georges, beyond such as Nature provided when she gave men fingers; and they burn attenuated

[17] M. Thury found this hole, and fathomed it to a depth of 6-1/2 mètres.

tallow candles with full-bodied wicks. Also, the tea is flavoured with vanille, unless that precious flavouring is omitted by private contract.

Chapter 3

The Lower Glacière of the Pré De S. Livres

I had intended to walk on from S. Georges to Bière, after returning from the glacière last described, and thence, the next morning, to the Pré de S. Livres, the mountain pasturage of the commune of S. Livres,[18] a village near Aubonne. But Renaud advised a change of plan, and the result showed that his advice was good. He said that the *fermier* of the Glacière of S. Livres generally lived in S. Georges, and, if he were at home, would be the best guide to the glacière; while the distance from S. Georges was, if anything, rather less than the distance from Bière; so that by remaining at the Cavalier for another night the walk to Bière would be saved, and the possibility of finding no competent guide there would be evaded. Jules Mignot, the farmer in question, was at home, and promised to go to the glacière in the morning, pledging his word and all that he was worth for the existence and soundness of the ladders; a matter of considerable

[18] *Sancti Liberii locus*, the Swiss Dryasdust explains. There is nothing to connect any known S. Liberius with this neighbourhood, unless it be the Armenian prince who secretly left his father's court for Jerusalem, and was sought for throughout Burgundy and other countries. It seems that Saint Oliver is merely a corruption of S. Liberius, the Italian form of the latter, Santo Liverio, having become Sant-Oliverio, as S. Otho became in another country Sant Odo, and thence San Todo, thus creating a new Saint, S. Todus.—Act SS. May 27.

importance, for M. Thury had been unable to reach the ice, as also my sisters, by reason of a failure in this respect.

In the course of the evening Mignot came in, and confidentially took the other chair. He wished to state that he had three *associés* in working the glacière, and that one of them knew of a similar cave, half an hour from the one more generally known; the *associé* had found it two years before, and had not seen it since, and he believed that no one else knew where it was to be found. If I cared to visit it, the *associé* would accompany us, but there was some particular reason--here he relapsed into patois--why this other man could not by himself serve as guide to both glacières. As this meant that I must have two guides, and suggested that perhaps the right rendering of *associé* was 'accomplice,' the negotiation nearly came to a violent end; but the farmer was so extremely explanatory and convincing, that I gave him another chance, asking him how much the two meant to have, and telling him that, although I could not see the necessity for two guides, I only wished to do what was right. He expressed his conviction of the truth of this statement with such fervour, that I could only hope his moderation might be as great as his faith. He took the usual five minutes to make up his mind what to say, going through abstruse calculations with a brow demonstratively bent, and, to all appearance, reckoning up exactly what was the least it could be done for, consistently with his duty to himself and his family. Then he asked, with an air of resignation, as if he were throwing himself and his *associé* away, 'Fifteen francs, then, would monsieur consider too much?' 'Certainly, far too much; twelve francs would be enormous. But, for the pleasure of his company and that of his friend, I should be happy to give that sum for the two, and they must feed themselves.' He jumped at the offer, with an alacrity which showed that I had much under-estimated his margin in putting it at three francs; and with many expressions of anticipatory gratitude, and promises of axes and ropes in case of emergency, he bowed himself out. The event proved that both the men were really valuable, and they got something over the six francs a-piece.

The rain had been steadily increasing in intensity for the last twenty-four hours, from the insidious steeping of a Scotch mist to the violence of a chronic thunderstorm, and had about reached this crisis when we started in the morning for the Pré de S. Livres. I had already tested its effects before breakfast, in a search for the Renaud of the day before, who had made statements regarding the ice at S. Georges, and the time of cutting it, which a night's reflection showed to be false. To search for Henri Renaud in the village of S. Georges, was something like making an enquiry of a certain porter for the rooms of Mr. John Jones. The landlady of the Cavalier was responsible for the first stage of the journey, asserting that he lived two doors beyond the next auberge, evidently with a feeling that it was wrong so far to patronise the rival house as to live near it. That, however, was not the same Henri Renaud; and a house a few yards off was recommended as a likely place, where, instead of Henri, a Louis Renaud turned up, shivering under the eaves in company with the *fermier*, who introduced Louis in due form as the accomplice. They received conjointly and submissively a lecture on the absurdity of calling it a rainy morning, and the impossibility of staying at home, even if it came on much worse, and then pointed the way to the true Henri Renaud, half-way down the village. When I arrived at the place indicated, and consulted a promiscuous Swiss as to the abode of the object of my search, he exclaimed, 'Henri Renaud? I am he.' 'But,' it was objected, 'it is the *marchand de bois* who is wanted.' 'Precisely, Henri Renaud, marchand de bois; it is I.' 'But, it is the cutter of ice in the glacière.' 'Ah, a different Henri. That Henri is in bed in the house yonder,' and so at last he was found. When finally unearthed, Henri confessed that when he had said *spring* the day before, he ought to have said *autumn*, and that by autumn he meant November and December. Enquiries elsewhere showed that the end of summer was what he really meant, if he meant to tell the truth.

Our route for the glacière followed the high road which leads by the Asile de Marchairuz to La Vallée, as far as the well-known Châlet de la S. Georges; and then the character of the way changed rapidly for the worse, and we took to the wet woods. After a time, the wood ceased for a while, and a large expanse of smooth rock showed itself, rising slightly from the horizontal, and so slippery in its present wet condition that we could not pass up it. Then woods again, and then the montagnes of *Sous la Roche*, and *La Foireuse*, till at last, in two hours, the Pré de S. Livres was achieved. The fog was so dense that nothing could be seen of the general lie of the country; but the *thalweg* was a sufficient guide, and after due perseverance we came upon the glacière, not many yards from that line, on the north slope of the open valley, about 4,500 feet above the sea.

To prevent cattle from falling into the pit, a wall has been built round the trees in which it lies. The circumference of this wall is 435 feet, but there are so many trees at the upper end of the enclosure that this gives an exaggerated idea of the size of the pit. The men fed while the preliminary measurements were being made; and when this was accomplished, they pressed their bottle of wine upon me so hospitably that I was obliged to antedate the result which its appearance promised, and plead *mal d'estomac*. Of all things, it is most unwise to give a reason for a negative, and so it proved in this instance; for they promptly felicitated themselves and me on the good luck by which it happened that they had brought a wine famous on all the côte as a remedy for that somewhat vague complaint--a homoeopathic remedy in allopathic doses.

The glacière is entered by a natural pit in the gentle slope of grass, not much unlike the pit of La Genollière, but wider, and covered at the bottom with snow.[19] The first ladder leads down to a ledge of rock on which bushes and trees grow, and this ledge it is possible to reach without a ladder; the next ladder leads on to the deep snow, and descent by any ordinary manner of climbing is in this case quite impossible.[20] The snow

[19] My sisters made a two-days' excursion from Arzier to this glacière in the autumn of 1862, and found no snow in the bottom of the pit. They took the route by Gimel to Bière, intending to defer the visit to the glacière to the morning of the second day; but being warned by the appearance known locally as *le sappeur qui fume*, a vaporous cloud at the mouth of a cavern near the Dent d'Oche, on the other side of the Lake of Geneva, they caught the communal forester at once, and put themselves under his guidance. The distance from Bière is two hours' good walking, and an hour and a half for the return. There was no ladder for the final descent, and the neighbouring châlet could provide nothing longer than 15 feet, the drop being 30 feet. Two Frenchmen had attempted to make their way to the cave a week before; but the old 30-foot ladder of the previous year broke under the foremost of them, and he fell into the pit, whence he was drawn up by means of a cord composed of rack-ropes from the châlet, tied together. However useful a string of cow-ties may be for rescuing a man from such a situation, A. and M. did not care to make use of that apparatus for a voluntary descent, so they were perforce contented with a distant view of the ice from the lower edge of the pit.

[20] See the section of this cave and pit.

slopes down towards a lofty arch in the rock which forms the north-west side of the pit, and this arch is the entrance to the glacière; it is 28-3/4 feet wide, and as soon as we passed under it we found that the snow became ice, and it was necessary to cut steps; for the surface of underground ice is so slippery, unlike the surface of ordinary glaciers, that the slightest defect from the horizontal makes the use of the axe advisable. The stream of ice falls gradually, spreading out laterally like a fan, so as to accommodate itself to the shape of the cave, which it fills up to the side walls; it increases in breadth from 28-3/4 feet at the top to 72 feet at the bottom of the slope, and the distance from the top of the first ladder to this point is 177 feet. Here we were arrested by a strange wall of ice 22 feet high, down which there seemed at first no means of passing; but finding an old ladder frozen into a part of the wall, we chopped out holes between the upper steps, and so descended, landing on a flooring composed of broken blocks and columns of ice, with a certain amount of what seemed to be drifted snow. This wall of ice, which was 72 feet long and 22 feet high, was not vertical, but sloped the wrong way, caving in under the stream of ice; and from the projecting top of the wall a long fringe of vast icicles hung down, along the whole breadth of the fan. The effect of this was, that we could walk between the ice-wall and the icicles as in a cloister, with solid ice on the one hand and Gothic arcades of ice on the other, the floor being likewise of ice, and the roof formed by the junction of the wall with the top of the icicle-arcade. The floor of this cloister was not 22 feet below the top of the wall, for it formed the upper part of a gentle descending slope of ice, rounded off like a fall of water, which seemed to flow from the lower part of the wall; and the height of 22 feet is reckoned from the foot of this slope, which terminated at a few feet of horizontal distance from the foot of the wall. The wall of ice was plainly marked with horizontal bands, corresponding, no doubt, to a number of years of successive deposits; sometimes a few leaves, but more generally a strip of minuter débris, signified the divisions between the annual layers. There had been many columns of ice from fissures in the rock, but all had fallen except one large ice-cascade, which flowed from a hole in the side of the cave on to the main stream, about two-thirds of the distance down from the snow. One particularly grand column had stood on the very edge of the ice-wall, and its remains now lay below.

The flooring of mingled ice and snow, on which we stood, sloped through about five vertical feet from the foot of the wall, and came to an end on broken rocks, from which the terminal wall of the cave sprang up. The effect of the view from this point, as we looked up the long slope of ice to where the ladders and a small piece of sky were visible, was most striking. The accompanying engraving is from a sketch which attempts to represent it; the reality is much less prim, and much more full of beautiful detail, but still the engraving gives a fair idea of the general appearance of the cave.

While I was occupied in making sketches and measurements, Mignot was engaged in chopping discontentedly at the floor, in two or three different places. At length he seemed to find a place to his mind, and chopped perseveringly till his axe went through, and then he suggested that we should follow. The hole was not tempting. It opened into the blackest possible darkness, and Mignot thrust his legs through, feeling for a foothold, which, by lowering himself almost to his armpits, he soon discovered: the foothold, however, proved to be a loose stone, which gave way under him and bounded down, apparently over an incline of like stones, to a distance which sounded very alarming. But he would not give in, and at length, descending still further by means of the snow in which the hole was made, he was rewarded by finding a solid block which bore his weight, and he speedily disappeared altogether, summoning me to follow. I proposed to light a candle first, not caring to go through such a hole, in such a floor, into no one knew what; but he was so very peremptory, evidently thinking that if he had gone through without a pioneering candle his monsieur might do the same, that there was nothing for it but to obey. The hole was very near the junction of the floor with the slope of stones where the floor terminated, and the space between the hole and the slope seemed to be filled up with a confused mass of snow and ice, in which the snow largely predominated; so that there was good hold for hands and feet in passing down to the stones, which might be about 7 feet below the upper surface of the floor.

Here we crouched in the darkness, with our faces turned away from the presumed slope of stones, till a light was struck. The accomplice did not find it in the bond that he should go down, and he preferred to reserve his energies for his own peculiar glacière.

Figure 4. Lower Glacière of the Pré De S. Livres.

As soon as the candle had mastered a portion of the darkness, we found that we were squatting on a steeply sloping descent of large blocks of stone, while in face of us was a magnificent wall of ice, evidently the continuation of the wall above, marked most plainly with horizontal lines. This wall passed down vertically to join the slope on which we were, at a depth below our feet which the light of the candle had not yet fathomed. The horizontal bands were so clear, that, if we had possessed climbing apparatus, we could have counted the number of layers with accuracy. Of course we scrambled down the stones, and found after a time that the angle formed by the ice-wall and the slope of stones was choked up at the bottom by large pieces of rock, one piled on another just as they had fallen from the higher parts. These blocks were so large, that we were able to get down among the interstices, in a spiral manner, for some little distance; and when we were finally stopped, still the ice-wall passed on below our feet, and there was no possible chance of determining to what depth it went. The atmosphere at this point was a sort of frozen vapour, most unpleasant in all respects, and the candles burned very dimly. The thermometer stood at 32°, half-way down the slope of stones.

We were able to stretch a string in a straight line from the lowest point we reached, through the interstices of the blocks of stone, and up to the entrance-hole, and this measurement gave 50 feet. Considering the inclination of the upper ice-floor, and the sharpness of the angle between the wall of ice and the line of our descent to this lowest point, I believe that 50 feet will fairly represent the height of the ice-wall from this point to the foot of the slope from the upper wall; so that 72 feet will be the whole depth of ice, from the top of the third ladder to the point where our further progress downwards was arrested. The correctness of this calculation depends upon the honesty of Mignot, who had charge of the farther end of the string, and was proud of the wonders of his cave.

A dishonest man might easily, under the circumstances, have pulled up a few feet more of string than was necessary, but 50 feet seemed in no way an improbable result of the measurement.

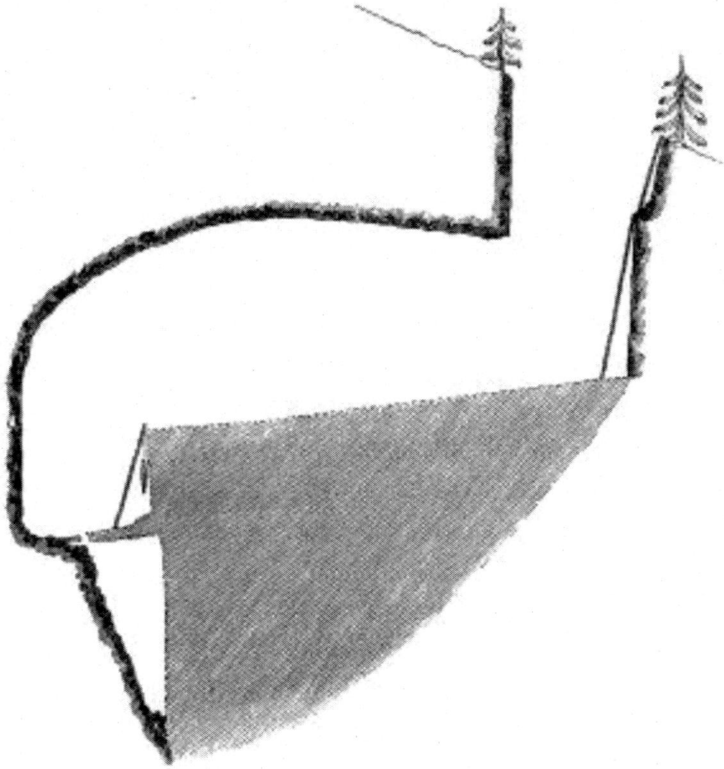

Figure 5. Section of the Lower Glacière of the Pré De S. Livres.

The ice was as solid and firm as can well be conceived. The horizontal bands would seem to prove conclusively that it was no coating of greater or less thickness on the face of a vertical wall of rock, an idea which might suggest itself to anyone who had not seen it, and I think it probable that the amount of ice represented in the section of the cave is not an exaggeration. We were unable to measure the whole length of the wall in the lower cave, from the large number of blocks of stone which had fallen at one end, and lay against its face. Probably, from the nature of the case, it was not so long as the 72 feet of wall above; but we measured 50 feet, and could see it still passing on to the right hand as we faced it. In trying to penetrate farther along the face, I found a wing of the brown fly we had seen in considerable abundance on the ice in La Genollière, frozen into the remains of a column.

There was so very much to be observed on all sides, and the measurements took up so much time, owing to the peculiar difficulties which attended them, that I did not examine with sufficient care the curious floor of ice through which we cut our way to the lower cavern. Neither did I notice the roof of the cavern thus reached, which may be very different from the shape of the upper surface of the floor composing it. If the ice-wall goes straight up, and the roof is formed of the ice-floor alone, then it is a very remarkable feature indeed. But, more probably, the lower wall leans over more and more towards the top, and so forms as it were a part of the roof. It is possible that, as the wall has grown, each successive annual layer has projected farther and farther, till at last some year very favourable to the increase of ice has carried the projection for that year nearly to the opposite stones, and then an unfavourable year or two would form the foot of the upper wall. This seems more probable, from the loose constitution of the floor at the point where it joins the stones, as if it were there only made up of drift and débris, while the part of the floor nearer the foot of the wall is solid ice. It has been suggested to me that possibly water accumulates in the time of greatest thaw to a very large extent in the lower parts of the cave, and the ice-floor is formed where the frost first takes hold of this water. But the slope of the ice-floor is against this theory, to a certain extent; and the amount of water necessary to fill the cavity would be so enormous, that it is contrary to all experience to imagine such a collection, especially as the cave showed no signs of present thaw. The appearance of the rocks, too, in the lower cave, and the surface of the ice-wall there, gave no indications of the action of water; and there was no trace of ice among the stones, as there certainly would have been if water had filled the cave, and gradually retired before the attacks of frost, or in consequence of the opening up of drainage. There were pieces of the trunks of trees, also, and large bones, lying about at different levels on the rocks. I never searched for bones in these caves, owing to the absence of the stalagmitic covering which preserves cavern-bones from decay; nor did I take any notice of such as presented themselves without search, for the *bergers* are in the habit of throwing the carcases of deceased cows into any deep hole in the neighbourhood of the place where the carcases may be found, in consequence of the general belief that living cows go mad if they find the grave of a companion; so that I should probably have made a laborious collection of the bones of the *bos domesticus*.

This belief of the bergers respecting the cows is supported by several circumstantial and apparently trustworthy accounts of fearful fights among herds of cattle over the grave of some of the herd. The sight of a companion's blood is said to have a similar effect upon them. Thus a small pasturage between Anzeindaz and the Col de Cheville, on the border of the cantons Vaud and Valais, is still called *Boulaire* from legendary times, when the herdsmen of Vaud (then Berne) won back from certain Valaisan thieves the cattle the latter were carrying off from La Varraz. Some of the cows were wounded in the battle, and the sight of their blood drove the others mad, so that they fought till almost all the herd was destroyed; whence the name Boulaire, from *ébouëler*, to disembowel,--a word formed from *bouë*, the patois for *boyau*.

When we left the lower darkness and ascended to the floor of ice once more, Mignot expressed a desire to see my attempt at a sketch of the glacière from that point, as he had been much struck during his negotiatory visit of the night before by the sketch of the entrance to the Glacière of S. Georges, chiefly because he had guessed what it was meant for. He was evidently disappointed with the representation of his own cave, for he could see nothing but a network of lines, with unintelligible words written here and there, and after some hesitation he confessed that it was not the least like it. A little explanation soon set that right, and then he began to plead vigorously for the wall which surrounded the trees at the mouth of the pit. Why was it not put in? He was told, because it could not be seen from below; but nevertheless he strongly urged its introduction, on the ground that he had built it himself, and it was such a well-built wall; facts which far more than balanced any little impossibility that might otherwise have prevented its appearance. After we had reached the grass of the outer world again, he made me sketch the entrance to the pit, pointing to the containing wall with parental pride, and standing over the sketch-book and the sketcher with an umbrella which speedily turned inside out under the combined pressure of wind, and rain, and years; a feat which it had already performed *des fois*, he said, in the course of his acquaintance with it.

Before finally leaving the glacière, I examined the structure of the great stream of ice, at different points near the top of the limiting wall. From its outward appearance it might have been expected to be rough, but it was not so; it was knotty to the eye, but perfectly smooth to the foot, and, when cut, showed itself perfectly clear and limpid. It did not separate under the axe into misshapen pieces, with faces of every possible variation from regularity, that is, with what is called vitreous fracture, but rather separated into a number of nuts of limpid ice, each being of a prismatic form, and of much regularity in shape and size. It was smooth, dark-grey, and clear; free from air, and free from surface lines; very hard, and suggesting the idea of coarse internal granulation. In the large ice-streams of some darker glacières, this ice assumed a rather lighter colour by candle-light, but always presented the same granular appearance, and cut up into the same prismatic nuts, and was evidently free from constitutional opacity.

Chapter 4

The Upper Glacière of the Pré De S. Livres

We now put ourselves under the guidance of the accomplice, Louis, who began to express doubts of his ability to find the upper glacière, administering consolation by reminding us that if he could not find it no one else could.

As we walked on through the mist and rain, it became necessary to circumvent a fierce-looking bull, and Mignot and the accomplice told rival tales of the dangers to which pedestrians are exposed from the violence of the cattle on some montagnes, where the bulls are allowed to grow to full size and fierceness. Mignot was quite motherly in his advice and his cautions, recommending as the surest safeguard a pocket-pistol, loaded with powder only, to be flashed in the bull's face as he makes his charge. When informed that in England an umbrella or a parasol is found to answer this purpose, he shook his head negatively, evidently having no confidence in his own umbrella, and doubting its obeying his wishes at the critical moment; indeed, it would require a considerable time, and much care and labour, to unfurl a lumbering instrument of that description. He had the best of the tale-contest with Renaud in the end, for he had himself been grazed by a bull which came up with him at the moment when he sprang into a tree.

Before very long we reached a little kennel-like hut of boughs, which no decent dog would have lived in, and no large dog could have entered, and from this we drew a charcoal-burner. No, he said, he did not know the glacière; he had heard that one had been discovered near there, and he had spent hours in searching for it without success. A herdsman on his way from one pasturage to another could give no better help, and we began to despair, till at length Louis desired us to halt in a place sheltered from the rain, while he prosecuted the search alone. We had abundant time for observing that, like other leafy places sheltered from the rain, our resting-place was commanded by huge and frequent drops of water; but at last a joyful *Jodel* announced the success of the accomplice, and we ran off to join him.

At first sight there was very little to see. Louis had lately been enunciating an opinion that the cave was not worth visiting, and I now felt inclined to agree with him. The general plan appeared to be much the same as in the one we had just left, but the scale was considerably smaller. The pit was not nearly so deep or so large, and, owing to the falling-in of rock and earth at one side, the snow was approached by a winding path with a gradual fall. As soon as the snow was reached, the slope became very steep, and led promptly to an arch in the rock, where the stream of ice began. The cave being shallow, the stream soon came to an end, and, unlike that in the lower glacière, it filled the cave down to the terminal wall, and did not fill it up to the left wall. Here the ground of the cave was visible, strewn with the remains of columns, and showing the thickness of the bottom of the stream to be about 6 feet only. The arch of entrance had evidently been almost closed by a succession of large columns, but these had succumbed to the rain and heat to which they had been exposed by their position.

The left side of the cave, in descending, that is the west side, was comparatively light, being in the line from the arch; but the other side was quite dark, and after a time we found that the ice-stream, instead of terminating as we had supposed with the wall of rock at the end of the cavern, turned off to the right, and was lost in the darkness. Of course candles were brought out, though Louis assured us that he had explored this part of the cave on his previous visit, and had found that the right wall of the cave very soon stopped the stream: we, on the contrary, by tying a candle to a long stick, and thrusting it down the slope of ice, found that the stream passed down extremely steeply, and poured under a narrow and low arch in the wall of the cave, beyond which nothing could be seen. We despatched pieces of ice along the slope, and could hear them whizzing on after they had passed the arch, and landing apparently on stones far below; so I called for the cords, and told Louis that we must cut our way down. But, alas! the cords had been left at the other glacière! One long bag, with a hole in the middle like an old-fashioned purse, had carried the luncheon at one end and the ropes at the other; and when the luncheon was finished, the bag had been stowed away under safe trees till our return. This was of course immensely annoying, and I rang the changes on the few words of abuse which invention or knowledge supplied, as we sat damp and shivering on the verge of the slope, idly sending down pieces of broken columns which brought forth tantalising sounds from the subterranean regions. At length Renaud was moved to shame, and declared that he would cut his way down, rope or no rope; but this seemed so horribly hazardous a proceeding under all the circumstances, that I forbad his attempting it. Seeing, however, that he was determined to do something, we arranged ourselves into an apparatus something like a sliding telescope. Louis cut a first step down the slope, and there took his stand till such time as Mignot got a firm grasp of the tail of his blouse with both hands, I meanwhile holding Mignot's tail with one hand, and the long stick with the candle attached to it with the other; thus professedly supporting the whole apparatus, and giving the necessary light for the work. Even so, we tried again to persuade Renaud to give it up, but he was warmed to his work, and really the arrangement answered remarkably well: when he wished to descend to a new step, Mignot let out a little blouse, and, being himself similarly relieved, descended likewise a step, and then the remaining link of the chain followed. The leader slipped once, but fortunately grasped a projecting piece of rock, for the stream was here confined within narrow walls, and so the strength of the apparatus was not tested; it could scarcely have stood any serious call upon its powers.

After a considerable period of very slow progress, Renaud asked for the candlestick, never more literally a stick than now, and thrust it under the arch, stooping down so as to see what the farther darkness might contain. We above could see nothing, but, after an anxious pause, he cried *On peut aller!* with a lively satisfaction so completely shared by Mignot, that that worthy person was on the point of letting Renaud's blouse go, in order to indulge in gestures of delight. The step-cutting went on merrily after this announcement, and one by one we came to the arch and passed through, finding it rather a trough than an arch; the breadth was about 4 feet, and the height from 1-1/4 to 1-1/2 feet, and, as we pushed through, our breasts were pressed on to the ice, while our backs scraped against the rock which formed the roof.

As soon as this trough was passed, the ice spread out like a fan, and finally landed us in a subterranean cavern, 72 feet long by 36 feet broad, to which this was the only entrance.

Figure 6. Second Cave Of The Upper Glacière Of The Pré De S. Livres.

The breadth of the fan at the bottom was 27 feet; and near the archway a very striking column poured from a vertical fissure in the wall, and joined the main stream. The fissure was partially open to the cave, and showed the solid round column within the rock: this column measured 18-1/2 feet in circumference, a little below the point where it became free of the fissure, and it had a stream of ice 22 feet long pouring from its base. The colour of the column was unusual, being a dull yellowish green, and the peculiar structure of the ice gave the whole mass the appearance of coursing down very rapidly, as if the water had been frozen while thus moving, and had not therefore ceased so to move. At the bottom of the fan, the flooring of the cave consisted of broken stones for a small space, and then came a black lake of ice, which occupied all the centre of the cave, and afforded us no opportunity of even guessing at its depth. From the manner, however, in which it blended with the stones at its edge, I am not inclined to believe that this depth was anything very great.

Renaud, in his impetuosity, had ceased to cut steps towards the bottom of the slope, and had slipped down the last few feet, of course cutting the remaining steps before attempting to reascend. We found him strutting about the floor of the cave, tossing his wet cap in the air, and crying *No one! No one! I the first!*, declining to take any part in measurements until the full of his delight and pride had been poured out. He shouted so loud that I was obliged to stop him, lest by some chance the unwonted disturbance of the air should bring down an unstable block from the roof of the arch, and seal us up for ever. There was no sign of incipient thaw in the cave, and the air was very dry, so much so as at once to call attention to the fact. At the farthest end, a lofty dome opened up in the roof; and possibly at some time or other the rock may here fall through, and afford another means of entrance. Beneath this dome a very lovely cluster of columns had grouped itself, formed of the clear porcelain-like ice, and fretted and festooned with the utmost delicacy, as if Andersen's Ice Maiden had been there in one of her amiable moods, and had built herself a palace. This dome in the roof was similar to many which I afterwards observed in other glacières, being a vertical fissure with flutings from top to bottom--not a spherical dome, but of that more elegant shape which the female dress of modern times assumes on a tall person.

Between the base of the circular column and the wall, we found a rare instance of clear jelly-like ice, without any lines external or internal, such as is formed in the open air under very favourable circumstances.

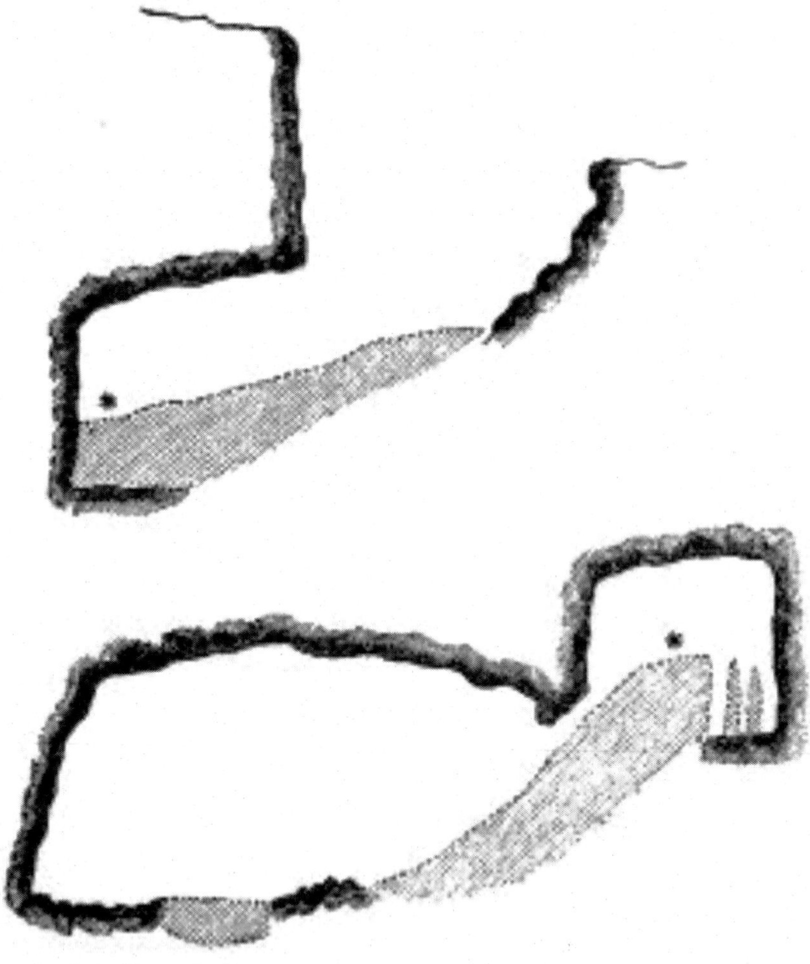

[21][marker is in illustration]

Figure 7. Vertical Sections of the Upper Glacière of the Pré De S. Livres.

[21] A point common to the two sections, which are made by planes nearly at right angles to each other.

The ordinary number of undergraduate May Terms had afforded various opportunities for studying the comparative clearness of different pieces of ice, but certainly no one ever saw a lemon pippin through an inch and a half of that material so clearly as we now saw the white rock through 1-1/2 feet. Mignot, indeed, said 2 feet; but it was his way to make a large estimate of dimensions, and he constantly interrupted my record of measurements by the assertion that I had made them *moins que plus*. We were all disappointed by the actual size of the ice-fall which it had cost us so much time and trouble to descend, the distance from the first step to the last being only 26 feet: as this, however, was given by a string stretched from the one point to the other, and not following the concave surface of the ice, the real distance was something more than this.

It was now getting rather late, considering the journey one of us had yet to perform, and we walked quickly away from the glacière, agreeing that it was not improbable that in that part of the Jura there might be many hidden caves containing more or less ice, with no entrance from the world outside, except the fissures which afford a way for the water. The entrance to this cave was so small, that the same physical effect might well be produced by one or two cracks in the rock, such as everyone is well acquainted with who has walked on the fissured limestone summits of the lower mountains; and, indeed, Renaud positively affirmed that at the time of his former visit there was not even this entrance to the lower cave, for the ice-stream reached then a higher point of the wall, and completely filled and hid the arch we had discovered. It is very difficult to see how ice can exist in a cave which has no atmospheric communication with the colds of winter, as would apparently be the case with this cave if the one entrance were closed; but where the cracks and small fissures in the rock do provide such communication, there is no reason why we should not imagine all manner of glacial beauties decorating unknown cavities, beyond the general physical law to which all the glacières would seem to be exceptions.

Mignot now became communicative as to the amount of ice supplied by his glacière, the lower of the two we had seen; and his statistics were so utterly confused, that I gave him ten centimes and an address, and charged him to write it all down from his account-book, and send it by post. The letter was accordingly written on July 24, and after trying many unsuccessful addresses in various parts of Switzerland, it finally reached England in the middle of September. It tells its own tale sufficiently well, and is therefore given here with all the mistakes of the original.

'Mon cher Monsieur Browne,--J'ai beaucoup tardé a vous écrire les détails promis, sans doute je ne voulait pas vous oublier; nous sommes affligés dans nôtre maison ma femme et gravement malade ce qui me donne beaucoup de tourment jour et nuit, enfin ce n'est pas ce qui doit faire nôtre entretient.

En 1863. Nous avons exploité comme suit. (Dépenses.)

Aoust	27	10 journées pour confectionner les Echelles et les poser.
"	29	3 journées pour couper la glasse.
"	31	11 journées pour sortir la glasse avec les hôtes.
"	31	4 chars a deux chevaux pour ammener
	Menés	la charge a deux: dès St. Georges a
		Gland plusieurs autres journées pour accompagner
	Septembre	les chars. 70 pots de vin bu
	1	en faisant ces chargements, pour trois
		cordes pour se tenir.
	Septembre 2	Trois journées pour couper.
	le 3	12 journées pour sortir.

'Cher Monsieur.--Je ne vous ait pas mis le prix de chaque articles; ni tout-a fait tous les traveaux mais pour vous donner une idée, je veux vous donner connaissance du coût général des dépences pour deux chargements s'élève a 535 francs. Je vous donne aussi connaissance de la quantité de glasse rendue 235 quinteaux a 3 francs, qui produit 705 francs reste net sur ces deux chargements 175 francs: par conséquent mon cher Monsieur je n'ai pas besoin de vous donner des détails des chargements suivants c'est a peu près les mêmes frais, et la quantité de glasse aussi.

'Nous en avons refait trois chargements:--

Un le 15 Septembre.

2 le 13 Octobre.

3 le 14 Novembre

'Cela comprend toute l'exploitation de 1863.
'Vous m'excuserez beaucoup de mon retard.

'Je termine en vous présentant mes respectueuses salutations. Vous noublierez pas ce que vous mavez promis'[22] St. Georges, le 24 Juillet, 1864. *Dimanche*.

'Jules Mignot'

Instead of three francs the quintal, Mignot had previously told me that he got four francs, delivered at Gland, and five at Geneva. His ordinary staff during the time of the exploitation was ten men to carry and load, and two to cut the ice in the cave.

It was a matter of considerable importance to catch the Poste at Gimel, and the two Swiss groaned loudly on the consequent pace, unnecessary, as far as they were concerned, for the Poste was nothing to them. As a general rule, the Swiss of this district cannot walk so fast as their Burgundian or French neighbours, unless it is very much to their interest to do so, and then they can go fast enough. A legend is still preserved in the valleys of Joux and Les Rousses, to the following effect. While the Franche Comté was still Spanish, in 1648, commissioners were appointed to fix the boundaries between Berne and Burgundy, on the other side of the range of hill we were now descending, and they decided that one of the boundary

[22] The dimensions of the two caves, and of the various masses of ice.

stones must be placed at the distance of a common league from the Lake of Les Rousses. Unfortunately, no one could say what a common league was, beyond the vague definition of 'an hour's walk;' so two men were started from the shore of the lake, the one a Burgundian and the other a Swiss, with directions to walk for an hour down the Orbe towards Chenit, the stone to be placed half-way between the points they should respectively reach at the end of the hour. It was for the interest of the Franche Comté that the stone should be as near the lake as possible, and accordingly the Swiss champion made such walking as had never been seen before, and gained for Berne a considerable amount of territory. There was no such tragic result in this case as that which induced the Carthaginians to pay divine honours to the brothers whose speed, on a like occasion, had added an appreciable amount to the possessions of the republic.

At length we reached the point where the roads for Gimel and S. Georges separate, and there, under a glorious sapin, we said our adieux, and wished our *au revoirs*, and settled those little matters which the best friends must settle, when one is of the nature of a monsieur, and the others are guides. They burdened their souls with many politenesses, and so we parted. The inclemency of the weather was such, that the people in the lower country asked, as they passed, whether snow had fallen in the mountains, and the cold rain continued unceasingly down to the large plain on which the Federal Camp of Bière[23] is placed. Here for a few moments the sun showed itself, lighting up the white tents, and displaying to great advantage the masses of scented orchises, and the feathery *reine-des-prés*, which hemmed the road in on either side. All through the earlier part of the day, flowers had forced themselves upon our notice as mere vehicles for collected rain, when we came in contact with them; but now, for a short time, they resumed their proper place,--only for a short time, for the rain soon returned, and did not cease till midnight. Not all the garden scenery about Aubonne and Allaman (*ad Lemannum*), nor all the vineyards which

[23] The Cartulary of Lausanne states that the wealthy village of Bière received its name from the following historical fact:—In 522, the Bishop of Lausanne, S. Prothais, was superintending the cutting of wood in the Jura for his cathedral, when he died suddenly, and was carried down on a litter to a place where a proper *bier* could he procured, whence the place was named Bière.

yield the choice white wine of the Côte, could counterbalance the united discomfort of the rain, and the cold which had got into the system in the two glacières; and matters were not mended by the discovery that *Bradshaw* was treacherous, and that a junction with dry baggage at Neufchâtel could not be effected before eleven at night.

There are some curious natural phenomena in this neighbourhood, due to the subterranean courses which the fissured limestone of the Jura affords to the meteoric waters. Not far from Bière, the river Aubonne springs out at the bottom of an amphitheatre of rock, receiving additions soon after from a group of twenty natural pits, which the peasants call unfathomable--an epithet freely applied to the strange holes found in the Jura. It is remarkable that the way seems to stand at different levels in the various pits.[24] The plain of Champagne, in which they occur, is unlike the surrounding soil in being formed of calcareous detritus, evidently brought down by some means or other from the Jura, and is dry and parched up to the very edges of the pits. The Toleure, a tributary of the Aubonne, frequently large enough to be called a confluent, flows out from the foot of a wall of rock composed of regular parallelopipeds, and in the spring, when the snows are melting freely, its sources burst out at various levels of the rock. Farther to the west, the Versoie, famous for its trout, pours forth a full-sized stream near the Château of Divonne, which is said to take its name (*Divorum unda*) from this phenomenon. Passing to the northern slope of this range of the Jura, the Orbe is a remarkable example of the same sort of thing, flowing out peacefully in very considerable bulk from an arch at the bottom of a perpendicular rock of great height. This river no

---

[24] The most curious pit of this kind is the *frais-puits* of Vesoul, in the Vosgian Jura, which pours forth immense quantities of water after rain has fallen in the neighbourhood. The water rushes out in the shape of a fountain, and on one occasion, in November 1557, saved the town of Vesoul from pillage by a passing army. This pit is carefully described by M. Hassenfratz, in the *Journal de Physique*, t. xx. p. 259 (an. 1782), where he says that Cæsar was driven away from the town of Vesoul, which he had intended to besiege, by the floods of water poured forth from the *frais-puits*. I know of no such incident in Cæsar's life, though M. Hassenfratz quotes Cæsar's own words: the town of Vesoul, too, had no historical existence before the 9th or 10th century of our era. There is also a pit near Vesoul which contains icicles in summer, and may be the same as the *frais-puits*, for the old historian of Franche Comté, Gollut, in describing the latter, mentions that it is so cold that no one cares to explore it (pp. 91. 92).

doubt owes its origin to the superfluous waters of the Lake of Brenets, which have no visible outlet, and sink into fissures and *entonnoirs* in the rock at the edge of the lake. Notwithstanding that the lake is three-quarters of a league distant, horizontally, and nearly 700 feet higher, the belief had always been that it was the source of the stream, and in 1776 this was proved to be the fact. For some years before that date, the waters of the Lake of Joux had been inconveniently high, and the people determined to clean out the *entonnoirs* and fissures of the Lake of Brenets, which is only separated from the Lake of Joux by a narrow tongue of land, in the expectation that the water would then pass away more freely. In order to reach the fissures, they dammed up the outlet of the upper into the lower lake; but the pressure on the embankment became too great, and the waters burst through with much violence, creating an immense disturbance in the lake; and the Orbe, which had always been perfectly clear, was troubled and muddy for some little time. The source of the Loue, near Pontarlier, is more striking than even that of the Orbe.[25]

[25] See p. 122.

Chapter 5

The Glacière of the Grâce-Dieu, or La Baume, Near Besançon

The grand and lovely scenery of the Val de Travers has at length been opened up for the ordinary tourist world, by the railway which connects Pontarlier with Neufchâtel. The beauties of the valley are an unfortunate preparation for the dull expanse of ugly France which greets the traveller passing north from the former town; but the country soon assumes a pleasanter aspect, and nothing can be more charming than the soft green slopes, dotted with the richest pines, which form the approach to the station of Boujeailles. It is impossible for the most careless traveller to avoid observing the ill effects produced upon the trees on the south side of the forest of Chaux, by the crowded and neglected state in which they have been left, and the wet state of the soil. The branches become covered with moss, which first kills them, and then breaks them off, so that many tall and tapering sapins point their heads to the sky with trunks wholly guiltless of branches; while in other cases, where decay has not yet gone so far, the branches wear the appearance of gigantic stags' horns, with the velvet; and when a number of these interlace, the mosses unite in large dark patches, giving a cedar-like air to the scene of ruin.

Up to this point, an elderly Frenchman in the carriage had been extremely offensive, from the evil odour of his Macintosh coat; but in answer to a remark upon the improvement which the railway would effect, by providing ventilation for the forest, he gave so much information on that subject, and gave it so pleasantly, and had evidently so good a knowledge of the topography of Franche Comté, that his coat speedily lost its smell, and we became excellent friends.

It is a tantalising thing to be whirled on a hot and dusty day through districts famous for their wines, the dust and heat standing out in more painful colours by contrast with the recollection of cooling draughts which other occasions have owed to such vineyards; though, after all, the true method of facing heat with success is to drink no wine. At any rate, the vineyards of Arbois must always be interesting, and if the stories of the Templars' orgies be true, we may be sure that the chapelry which they possessed in that town would be a favourable place of residence with the order; possibly Rule XVI might there be somewhat relaxed. 'The good wine of Arbois,' *la meilleure cave de Bourgougne*, a judicious old writer says, had free entry into all the towns of the Comté; and when Burgundy was becoming imperial, Maximilian extended this privilege through all the towns of the empire. A hundred years later, it had so high a character, that the troops of Henri IV. turned away from the town, announcing that they did not wish to attack *ceulx estoient du naturel de leur vin, qui frappe partout*;[26] and the king was forced to come himself, with his constable and marshals, to beat down the walls, in the course of which undertaking his men felt the vigour of the inhabitants to a greater extent than he liked. It is said that when he had taken the town, the municipality received him in state, and supplied him with wine of the country. He praised the wine very highly, on which one of the body had the ill taste to assure him that they had a better wine than that. 'You keep it, perhaps,' was the royal rebuke, 'for a better occasion.' Henry had a great opinion of this wine; and the Duc de Sully states, in his Memoirs, that when the Duc de Mayenne retired from the league against the king, and came to Monceaux to tender his allegiance, Henry punished him for past offences by walking so fast about the grounds of the château, that the poor duke, what with his sciatica, and

[26] Jean Bontemps, Conseiller au bailliage d'Arbois.

what with his fat, at last told him with an expressive gesture that a minute more of it would kill him. The king thereupon let him go, and promised him some *vin d'Arbois* to set him right again.[27]

The present appearance of the town, as seen from the high level followed by the railway, scarcely recalls the time when Arbois was known as *le jardin de noblesse*, and Barbarossa dated thence his charters, or Jean Sans-peur held there the States of Burgundy. Gollut[28] tells a story of a dowager of Arbois, mother-in-law to Philip V. and Charles IV of France, which outdoes legend of Bishop Hatto. Mahaut d'Artois was an elderly lady remarkable for her charities, and was by consequence always surrounded by large crowds of poor folk during her residence at the Châtelaine, the ruins of which lie a mile or two from Arbois. On the occasion of a severe famine in Burgundy, she collected a band of her mendicant friends in a stable, and burned them all, saying that '*par pitié elle hauoit faict cela, considerant les peines que ces pauvres debuoient endurer en temps de si grande et tant estrange famine*.'

There is a Val d'Amour near Arbois, but the more beautiful valley of that name lies between Dôle and Besançon, and, as we passed its neighbourhood, my friend with the Macintosh informed me that as it was clear from my questions that I was drawing up a history of the Franche Comté, he must beg me to insert a legend respecting the origin of this name, Val d'Amour, which, he believed, had never appeared in print. I disclaimed the history, but accepted the legend, and here it is:--The Seigneur of Chissey was to marry the heiress of a neighbouring seigneurie, and, it is needless to add, she was very lovely, and he was handsome and brave. A lake separated the two châteaux, and the young man not unfrequently returned by water rather late in the evening; and so it fell out that one night he was drowned. The lady naturally grieved sorely for her loss, and put in train all possible means for recovering her lover's body. Time, however, passed on, and no success attended her efforts, till at length she caused the hills which dammed up the waters to be pierced, and

---

[27] 'Allez vous en reposer, rafraischir et boire un coup au chasteau, car vous en avez bon besoin; j'ay du vin d'Arbois en mes offices, dont je vous envoyeray deux bouteilles, car je scay bien que vous ne le hayés pas.'—*Petitot*. iii. 9.

[28] Mém. de la Comté de Bourgougne, Dôle, 1592, p. 486.

then De Chissey was found. A village sprang up near the outlet thus made, and took thence its name Percée, or, as men now spell it, Parcey; and the rich vegetation which speedily covered the valley, where once the lake had been, gave it such an air of happiness and beauty, that the people remembered its origin, and called it the Valley of Love. It is a fact that Parcy was not always so spelled, for Noble Constantin Thiehault, Sieur de Perrecey, was a witness to the treaty for the transference of a miraculous host from Faverney to Dôle in 1608, and old maps and books give it as Perrecey and Parrecey indifferently. The De Chisseys, whose names may be found among the female prebends of Château-Chalon, with its necessary sixteen quarters, filled a considerable place in the history of the Comté from the Crusades downwards, and known as *les Fols de Chissey*, the brave[29] and dashing, and witty De Chisseys--qualities which no doubt were possessed by the poor young man for whom the fair Chatelaine drained the Val d'Amour.

As we drew nearer to Besançon, each turn of the small streams, and each low rounded hill, might have served as an illustration to Cæsar's 'Commentaries.' Now at length it was seen how, whatever the result of a battle, there was always a *proximus collis* for the conquered party to retire to; and it would have been easy to find many suitable scenes for the critical engagement, where the woods sloped down to a strip of grass-land between their foot and the stream.

The Frenchman knew his Cæsar, but he put that general in the fourth century B.C. He made mistakes, too, in quoting him, which were easily detected by a memory bristling with the details of his phraseology, the indelible result of extracting the principal parts of his verbs, and the nominatives of his irregular nouns, from half a dozen generations of small boys. He promised me a rich Julian feast in Besançon, and was greatly affected when he found that the Englishman could give him Cæsar's description of his native town. He wholly denied the amphitheatre with which one of our handbooks has gifted it; and this denial was afterwards echoed by every one in Besançon, some even thinking it necessary to

[29] One of the Seigneurs de Chissey, Michaud de Changey, who died in high office in 1480, was known by preeminence as *le Brave*.

explain the difference between an amphitheatre and an arch of triumph, the latter still existing in the town. The Jesuit Dunod relates that the amphitheatre was to be seen at the beginning of the seventeenth century, in the ruined state in which the Alans and Vandals had left it after their successful siege in 406. It seems to have stood near the present site of the Madeleine.

It was a great satisfaction to find that the Frenchman had himself visited the glacière which was the object of my search, and was able to give some idea as to the manner of reaching it, for my information on the subject was confined to a vague notice that there was an ice-cave five leagues from Besançon. As so often happened in other cases, he advised me not to go to it, but rather, if I must see a cave, to go to the Grotto of Ocelles,[30] a collection of thirty or more caverns and galleries near the Doubs, below Besançon. Seeing, however, that I was bent on visiting the glacière, he advised me not to go on Sunday, for the Cardinal Archbishop had ordered the Trappists at the Chartreuse near not to receive guests on that day; while Saturday, he thought, was almost as bad, for nothing better than an omelette could be obtained on days of abstinence. Saturday, then, was clearly the day to be chosen.

The first sight of Besançon explains at once why Cæsar was so anxious to forestall Ariovistus by occupying Vesontio, although the hill on which the citadel stands is not so striking as the similar hill at Salins, and the engines of modern warfare would promptly print their telegrams on every stone and man in the place, from the neighbouring heights. The French Government has wisely taken warning from the bombardment by the Allies, and has covered the heights which command it on either side with friendly fortifications, in which lie the keys of the place. Historically, Besançon is a place of great interest. It witnessed the catastrophe of Julius Vindex, who had made terms with Rufus, the general sent against him by Nero, but was attacked by the troops of Rufus before they learned the

---

[30] Dr. Buckland visited these caves in 1826, to look for bones, of which he found a great number. Gollut (in 1592) spelled the name *Aucelle*, and derived it from *Auricella*, believing that the Romans worked a gold mine there. It is certain that both the Doubs and the Loue supplied very fine gold, and the Seigneurs of Longwy had a chain made of the gold of those rivers, which weighed 160 crowns.

alliance concluded between the two generals. Vindex was so much grieved by the slaughter of his troops, and the blow thus struck, by an unhappy accident, at his designs against the emperor, that he put himself to death at the gates of the town, while the fight was still going on.[31] The Bisuntians claim to themselves the glory acquired by the Sequani, whose chief city Vesontio was, by the overthrow of Julius Sabinus, who asserted that he was the grandson of a son of Julius Cæsar, and proclaimed himself emperor in the time of Vespasian. The Sequani proceeded against him of their own accord, and conquered him in the interest of the reigning emperor; and he and his wife Peponilla lived hid in a tomb for nine years. Here two sons were born to them; and when they were all discovered and carried to Rome, Peponilla prettily told the emperor that she had brought up two sons in the tomb, in order that there might be other voices to intercede for her husband's life besides her own. They were, however, put to death.[32]

To judge from the style of the hotels, Besançon is not visited by many English travellers; and yet it well repays a visit, providing those who care for such things with a full average of vaulted passages, and feudal gateways, and arcaded court-yards, with much less than the average of evil smell. There are gates of all shapes and times--Louis-Quatorze towers, and fortifications specially constructed under Vauban's own eye; while the approach to the town, from the land side, is by a tunnel, cut through the live rock which forms a solid chord to the arc described by the course of the river Doubs. This excavation, called appropriately the *Porte Taillée*, is attributed by the various inhabitants to pretty nearly all the famous emperors and kings who have lived from Julius Cæsar to Louis XIV.: it owes its origin, no doubt, to the construction of the aqueduct which formerly brought into the town the waters pouring out of the rock at Arcier, two leagues from Besançon, and was the work probably of M. Aurelius and L. Verus. Local antiquaries assign the aqueduct to Agrippa, the son-in-law of Augustus, apparently for no better reason than because he built a

---

[31] Dion Cass. lib. lxiii.

[32] Ib. lib. lxvi.

similar work in Rome. The arch of triumph[33] at the entrance to the upper town has been an inexhaustible subject of controversy for many generations of antiquaries, and up to the time of Dunod was generally attributed to Aurelian: that historian, however, believed that its sculptures represented the education of Crispus, the son of Constantine, and that the name Chrysopolis, by which Besançon was very generally known in early times, was only a corruption of Crispopolis. Earlier writers are in favour of the natural derivation of Chrysopolis, and assert that when the Senones lost their famous chief, the Brennus of Roman history, before Delphos, they built a town where Byzantium afterwards stood, and called it Bisantium and Chrysopolis, in memory of their city of those names at home.

The Hôtel du Nord is a rambling old house, comfortable after French ideas of comfort, and rejoicing in an excellent cuisine; though it is true that on one occasion, at least, *haricots verts à l'Anglaise* meant a mass of fibrous greens, swimming in a most un-English sea of artificial fat. It is a good place for studying the natural manners of the untravelled Frenchman, who there sits patiently at the table, for many minutes before dinner is served, with his napkin tucked in round his neck, and his countenance composed into a look of much resignation. The waiters are for the most part shock-headed boys, in angular-tail coats well up in the back of the neck, who frankly confess, when any order out of the common run of orders is given, that a German patois from the left bank of the Rhine is their only extensive language. One of these won my eternal gratitude by providing a clean fork at a crisis between the last savouries and the *plat doux*; for the usual practice with the waiters, when anyone neglected to secure his knife and fork for the next course, was to slip the plate from under the unwonted charge, and leave those instruments sprawling on the tablecloth in a vengeful mess of gravy. Chickens' bones were there dealt with on all sides as nature perhaps intended that they should be dealt with, namely, by taking them between finger and thumb, and removing superfluities with the teeth; and French officers with wasp-like waists, and red trousers gathered in plaits to match, boldly despised the sophistication

[33] Known locally as the *Porte Noire*, like the great *Porta Nigra* at Treves, and other Roman gates in Gaul.

than the average thickness. Even where to all appearance there was nothing but mud and earth, an unexpected fall or two showed that all was ice below. Whether the driver had previously experienced the treacherousness of this slope of ice, or whatever his motive might be, he left me to enter and explore alone.

The roof of the entrance is at first a mere shell, formed by the thin crust of rock on which the surface-earth and trees rest high overhead; but this rapidly becomes thicker, as shown in the section of the cave, and thus a sort of outer cave is formed, the real portal of the glacière being reached about 60 feet above the bottom of the slope. This outer cave presents a curious appearance, from the distinctness with which the several strata of the limestone are marked, the lower strata weathered and rounded off like the seats of an amphitheatre of the giants, and all, up to the shell-like roof, arranged in horizontal semicircles of various graduated sizes, showing their concavity; while at the bottom of the whole is seen a patch of darkness, with two masses of ice in its centre, looming out like grey ghosts at midnight. This darkness is of course the inner cave, the entrance to which, though it seems so small from above, is 78 feet broad.

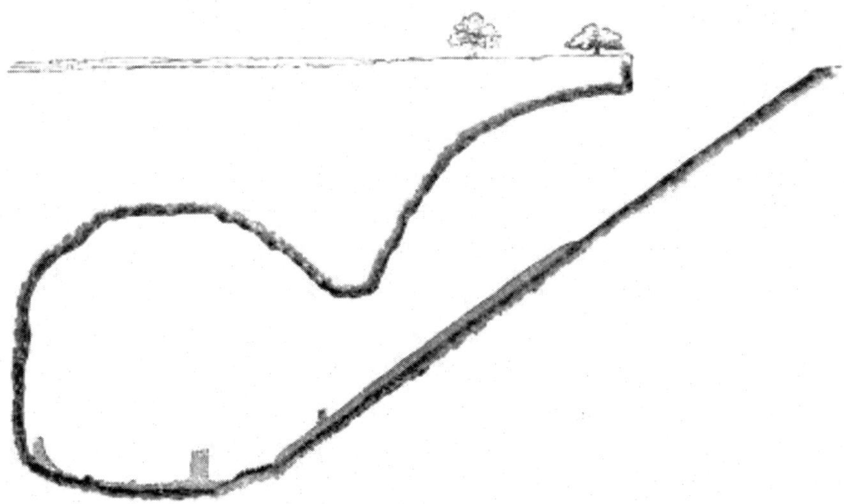

Figure 8. Vertical Section of the Glacière of Grâce-Dieu, Near Besançon.

The glacière itself may be said to commence as soon as this entrance, or perpendicular portal, is passed, and thus includes 60 feet of the long slope of ice, from the foot of which to the farther end of the cave is 145 feet, the greatest breadth of the cave being 148 feet.

Immediately below the portal I found a piece of the trunk of a large column of ice, 7 feet long and 12 feet in girth, its fractured ends giving the idea of the interior of a quickly-grown tree, in consequence of the concentric arrangement of convergent prisms described in the account of the Glacière of S. Georges. The wife of the farmer told me afterwards that there had been two glorious columns at this portal, which the recent rains had swept away. Excepting a short space at the foot of the slope, and another towards the farther end of the cave, the floor was covered with ice, in some parts from 3 to 4 feet thick: of this a considerable area had been removed to a depth of 2 1/2 or 3 feet, leaving a pond of water a foot deep, with bottom and banks of ice. The rock which composes the true floor rises at the farthest end of the cave, and the roof is so arranged that a sort of private chapel is there formed; and from a fissure in the dome a monster column of ice had been constructed on the floor, which, at the time of my visit, had lost its upper parts, and stood as a hollow truncated cone with sides a foot thick, and with seas of ice streaming from it, and covering the rising pavement of the chapel. Without an axe, and without help, I was unable to measure the girth of this column, which had not been without companions on a smaller scale in the immediate neighbourhood. At the west end of the cave, the wall was thickly covered for a large space with small limestone stalactites, producing the effect of many tiers of fringe on a shawl; while from a dark fissure in the roof a large piece of fluted drapery of the same material hung, calling to mind some of the vastly grander details of the grottoes of Hans-sur-Lesse in Belgium: down this wall there was also a long row of icicles, on the edges of a narrow fissure. The north-west corner was very dark, and an opening in the wall of rock high above the ground suggested a tantalising cave up there: the ground in this corner was occupied by the shattered remains of numerous columns of ice, which had originally covered a circular area between 60 and 70 feet in circumference.

The three large masses of ice which rendered this glacière in some respects more remarkable than any of those I have seen, lay in a line from east to west, across the middle of the cave, on that part of the floor where the ice was thickest. The central mass was extremely solid, but somewhat unmeaning in shape, being a rough irregular pyramid; its size alone, however, was sufficient to make it very striking, the girth being 66-1/2 feet at some distance from the ice-floor with which it blended. The mass which lay to the east of this was very lovely, owing to the good taste of someone who had found that much ice was wont to accumulate on that spot, and had accordingly fixed the trunk of a small fir-tree, with the upper branches complete, to receive the water from the corresponding fissure in the roof. The consequence was, that, while the actual tree had vanished from sight under its icy covering, excepting on one side where a slight investigation betrayed its presence, the mass of ice showed every possible fantasy of form which a mould so graceful could suggest. At the base, it was solid, with a circumference of 37 feet. The huge column, which had collected round the trunk of the fir-tree, branched out at the top into all varieties of eccentricity and beauty, each twig of the different boughs becoming, to all appearance, a solid bar of frosted ice, with graceful curve, affording a point of suspension for complicated groups of icicles, which streamed down side by side with emulous loveliness. In some of the recesses of the column, the ice assumed a pale blue colour; but as a rule it was white and very hard, not so regularly prismatic as the ice described in former glacières, but palpably crystalline, showing a structure not unlike granite, with a bold grain, and with a large predominance of the glittering element. But the westernmost mass was the grandest and most beautiful of all. It consisted of two lofty heads, like weeping willows in Carrara marble, with three or four others less lofty, resembling a family group of lions' heads in a subdued attitude of grief, richly decked with icy manes. Similar heads seemed to grow out here and there from the solid sides of the huge mass. The girth was 76 1/2 feet, measured about 2 feet from the floor. When this column was looked at from the side removed from the entrance to the cave, so that it stood in the centre of the light which poured down the long slope from the outer world, the transparency of the ice brought it to pass that the whole seemed set in a narrow frame of impalpable liquid blue, the effect of light penetrating through the mass at its extreme edges. The only means of determining the height of this column was by tying a stone to the end of a string, and lodging it on the highest head; but this was not an easy process, as I was naturally anxious not to injure the delicate beauty which made that head one of the loveliest things conceivable; and each careful essay with the stone seemed to involve as much responsibility as taking a shot at a hostile wicket, in a crisis of the game, instead of returning the ball in the conventional manner. When at last it was safely lodged, the height proved to be 27 feet. I had hoped to find it much more than this, from the grandeur of the effect of the whole mass, and I took the trouble to measure the knotted string again with a tape, to make sure that there was no mistake. The column formed upon the fir-tree was 3 or 4 feet lower.

I have since found many notices of this glacière in the Memoirs of the French Academy and elsewhere, extracts from which will be found in a later chapter. These accounts are spread over a period of 200 years, extending from 1590 to 1790, and almost all make mention of the columns or groups of columns I have described; but, without exception, the heights given or suggested in the various accounts are much less than those which I obtained as the result of careful measurement. The latest description of a visit to the glacière states a fact which probably will be held to explain, the present excess of height above that of earlier times.[37] The citizen Girod-Chantrans, who wrote this description, had procured the notes of a medical man living in the neighbourhood, from which it seemed that Dr. Oudot made the experiment, in 1779, of fixing stakes of wood in the heads of the columns, then from 4 to 5 feet high, and found that these stakes were the cause of a very large increase in the height of the columns, ice gathering round them in pillars a foot thick. So that it is not improbable that the largest of the three masses of the present day owes its height, and its peculiar form, to a series of stakes fixed from time to time in the various heads formed under the fissures in the roof, though nothing but the most solid ice can now be seen. It would be very interesting to try this

---

[37] Journal des Mines, Prairial, an iv., pp. 65, &c.

experiment in one of the caves where, without any artificial help, such immense masses of ice are formed; and by this means columns might, in the course of a year or two, be raised to the very roof. Further details on this subject will be given hereafter.

There was no perceptible draught of air in any part of the cave, and the candles burned steadily through the whole time of my visit, which occupied more than two hours. The centre was sufficiently lighted by the day; but in the western corner, and behind the largest column, artificial light was necessary. The ice itself did not generally show signs of thawing, but the whole cave was in a state of wetness, which made the process of measuring and investigating anything but pleasant. I had placed two thermometers at different points on my first entrance--one on a drawing-board on a large stone in the middle of the pond of water which has been mentioned, and the other on a bundle of pencils at the entrance of the end chapel, in a part of the cave where the ice-floor ceased for a while, and left the stones and rock bare. The former gave 33°, the latter, till I was on the point of leaving, 31 1/2°, when it fell suddenly to 31°. It was impossible, however, to stay any longer for the sake of watching the thermometer fall lower and lower below the freezing point; indeed, the results of sundry incautious fathomings of the various pools of water, and incessant contact of hands and feet with the ice, had already become so unpleasant, that I was obliged to desert my trusty hundred feet of string, and leave it lying on the ice, from want of finger-power to roll it up. The thermometers were both Casella's, but that which registered 31° was the more lively of the two, the other being mercurial, with a much thicker stem: the difference in sensitiveness was so great, that when they were equally exposed to the sun in driving home, the one ran up to 93° before the other had reached 85°.

In leaving the glacière, I found a little pathway turning off along the face of the rock on the left hand, a short way up the slope of entrance, and looking as if it might lead to the opening in the dark wall on the western side of the cave. After a time, however, it came to a corner which it seemed an unnecessary risk to attempt to pass alone; and my prudence was rewarded by the discovery that, after all, the supposed cave could not be thus reached. It is said that this other cave was the place to which the inhabitants fled for refuge when their district was invaded, probably by the Duke of Saxe-Weimar with his 10,000 Swedes, and that a ladder 40 feet long is necessary for getting at it.

The driver had long ago absconded when I returned to the upper regions; but the wife of the farmer of the grotto was there, and communicated all that she knew of the statistics of the ice annually removed. She said that in 1863 two chars were loaded every day for two months, each char taking about 600 kilos, the wholesale price in Besançon being 5 francs the hundred kilos. Since the quintal contains 50 kilos, it will be seen that this account does not agree with the statement of Renaud as to the amount of ice each char could take. No doubt, a char at S. Georges may mean one thing, and a char in the village of Chaux another; but the difference between 12 quintaux and 50 or 60 is too great to be thus explained, and probably Madame Briot made some mistake. Her husband, Louis Briot, works alone in the cave, and has twelve men and a donkey to carry the ice he quarries to the village of Chaux, a mile from the glacière, where it is loaded for conveyance to Besançon. He uses gunpowder for the flooring of ice, and expects the eighth part of a pound to blow out a cubic metre; and if, by ill luck, the ice thus procured has stones on the lower side, he has to saw off the bottom layer. Madame Briot said I was right in supposing March to be the great time for the formation of ice, as she had heard her husband say that the columns were higher then than at any other time of the year: she also confirmed my views as to the disastrous effects of heavy rain. As with every other glacière of which I could obtain any account, excepting the Lower Glacière of the Pré de S. Livres, she complained that the ice had not been so beautiful and so abundant this year as last, although the winter had been exceptionally severe.

Chapter 6

Besançon and Dôle

The afternoon was so far advanced when I returned to the convent, that it was clearly impossible to reach Besançon at five o'clock, and consequently there was time to inspect the Brothers and their buildings. The field near the convent was gay with haymakers; and the brown monks, with here and there a priest in *ci-devant* white, moved among the hired labourers, and stirred them up by exhortation and example,--with this difference, that while it was evidently the business of the monks so to do, the priests, on the other hand, had only taken fork in hand for the sake of a little gentle exercise. One unhappy Jacques Bonhomme made hot and toilsome hay in thick brown clothes, plainly manufactured from a defunct Brother's gown; for, to judge from appearances, a cast-off gown is a thing unknown. It was good to see a Brother, in horn spectacles of mediæval cut, tenderly chopping a log for firewood, and peering at it through his spectacles after each stroke, as a man examines some delicate piece of natural machinery with a microscope; to see another Brother, the sphere of whose duties lay in the flour-mill, standing in the doorway with brown robe and shaven crown all powdered alike with white, and a third covered from head to foot with sawdust; or, best of all, to see an antique Brother, with scarecrow legs, and low shoes which had presumably been in his possession or that of his predecessors for a long series of years, wheeling a barrow of liquid manure, with his gown looped up high by means of stout whipcord and an arrangement of large brass rings. The Brother whose business it was to do such cooking as might be required by visitors, grinned in the most friendly and engaging manner from ear to ear when he was looked at; and, by fixing him steadily with the eye, he could be kept for considerable spaces of time standing in the middle of the kitchen, knife in hand, with the corners of his mouth out of sight round his broad cheeks. His ample front was decked with a blue apron, suspended from his shoulders, and confined round the convexity of his waist by an old strap which no respectable costermonger would have used as harness. The soup served was by courtesy called *soupe maigre,* but it was in fact *soupe maigre* diluted by many homoeopathic myriads, and the Brother showed much curiosity as to my opinion of its taste--a curiosity which I could not satisfy without hurting his professional pride. When that course was finished, the large-faced cook suggested an omelette, as the most substantial thing allowed on eves, proceeding to draw the materials from a closet which so fully shared in the general abstinence from water as a means of cleansing, that I shut my eyes upon all further operations, and ate the eventual omelette in faith. Its excellence called forth such hearty commendations, that there seemed to be some danger of the mouth not coming right again. Then salads, and bread and butter, and wine, and various kinds of cheese were brought, which made in all a very fair dinner for a fast-day.

The culinary monk knew nothing of the history of his convent, beyond the bare year of its foundation, and displayed a monotonous dead level of ignorance on all topographical and historical questions: to him the *Pain d'Abbaye*[38] meant nothing further than the staff of life there provided, and

---

[38] One of the rights of the sovereigns of Burgundy was known by this name. The sovereign had the power of sending one soldier incapacitated by war to each abbey in the County, and the authorities of the abbey were bound to make him a prebendary for life. In 1602, after the siege of Ostend, the Archduke Albert exercised this right in favour of his wounded soldiers, forcing lay-prebendaries upon almost all the abbeys of the County of Burgundy. The Archduchess Isabella attempted to quarter such a prebendary upon the Abbey of Migette, a house of nuns, but the inmates successfully refused to receive the warrior among them (Dunod, *Hist. de l'Église de Besançon*, i. 367). For the similar right in the kingdom of France, see Pasquier, *Recherches de la France*, l. xii. p. 37. Louis XIV. did not exercise this

he neither knew himself nor could recommend any Brother who knew anything about the glacière. He was a German, and we talked of his native Baiern and the modern glories of his capital; and when his questions elicited a declaration of my profession, he passed up to Saxony, and pinned me with Luther. Finding that I objected to being so pinned, and repudiated something of that which his charge involved, he waived Luther, of whom he knew nothing beyond his name, and came down upon me triumphantly with the word Protestant. I explained to him, of course, that the worthy Elector, and his friends who protested, had not much to do with the Anglican branch of the Church Catholic; and then the old task had to be gone through of assuring the assembled Brothers that we in England have Sacraments, have Orders, have a Trinitarian Creed.

At length, about half-past three, we started for Besançon, paying of course *à volonté* for food and entertainment, as we did not choose to qualify as paupers. The driver told me on the way that there was another glacière at Vaise, a village three or four kilomètres from Besançon, and at no great distance from the road by which we should approach the town; so, when we reached the crest above Morre, where the road passes the final ridge by means of a tunnel, I paid the carriage off, and walked to the village of Vaise. The public-house knew of the glacière--knew indeed of two,--further still, kept the keys of both. This was good news, though the idea of keys in connection with an ice-cave was rather strange; and I proposed to organise an expedition at once to the glacières. The male half of the auberge declared that he was forbidden to open them to strangers, except by special order from a certain monsieur in Besançon; but the female half, scenting centimes, stated her belief that the monsieur in Besançon could never wish them to turn away a stranger who had come so many kilomètres through the dust to see the ice. She put the proposed disobedience in so persuasive and Christian a form, that I was obliged to take the husband's side,--not that he was in any need of support, for he had been longer married than Adam was, and showed no signs of giving way. It turned out, after all, that though there was no doubt about the existence

---

right after his conquest of the Franche Comté, perhaps because the Hôtel des Invalides, to which the Church was so large a contributor, met all his wants.

of the glacières, there was equally no doubt that they were *glacières artificielles*, being simply ice-houses dug in the side of a hill, and the property of a *glacier* in Besançon; so that my friend the driver had sent me to a mare's-nest.

The pathway across the hills to Besançon was rather intricate, and by good fortune an old Frenchman appeared, who was returning from his work at a neighbouring church, and served as companion and guide. He had bid farewell to sixty some years before, and, being a builder, had been going up and down a ladder all day, with full and empty *hottes*, to an extent which outdid the Shanars of missionary meetings; and yet he walked faster than any foreigner of my experience. He talked in due proportion, and told some interesting details of the bombardment of Besançon, which he remembered well. When he learned that I was not German, but English, he told me they did not say *Anglais* there, but *Gaudin*,--I was a *Gaudin*. This he repeated persistently many times, with an air worthy of General Cyrus Choke, and half convinced me that there was something in it, and that I might after all be a Gaudin. It was not till some hours after, that I remembered the indelible impression made by the piety of speech of recent generations of Englishmen upon the French nation at large, and thus was enabled to trace the origin of the name *Gaudin*. The old man evidently believed that it was the proper thing to call an Englishman by that name; thus reminding me of a story told of a French soldier in the Austrian service during the long early wars with Switzerland. The Austrians called the Swiss, in derision, Kühmelkers--a term more opprobrious than *bouviers*; and it is said that, after the battle of Frastens--one of the battles of the Suabian war,--a Frenchman threw himself at the feet of some Grisons soldiers, and innocently prayed thus for quarter; '*Très-chers, très-honorables, et très-dignes Kühmelkers! au nom de Dieu, ne me tuez pas*!'

The town of Besançon seems to spend its Sunday in fishing, and is apparently well contented with that very limited success which is wont to attend a Frenchman's efforts in this branch of *le sport*. There is a proverb in the patois of Vaud which says '*Kan on vau dau pesson, sé fo molli*;'[39] and

[39] '*Quand on veut du poisson, il se faut mouiller*;' referring probably to the method of taking trout practised in the Ormont valley, the habitat of the purest form of the patois. A man wades in

on this the Bisuntians act, standing patiently half-way up the thigh in the river, as the Swiss on the Lake of Geneva and other lakes may be seen to do. It is all very well to wade for a good salmon cast, or to spend some hours in a swift-foot[40] Scotch stream for the sake of a lively basket of trout; but to stand in a Sunday coat and hat, and 2-1/2 feet of water, watching a large bung hopelessly unmoved on the surface, is a thing reserved for a Frenchman indulging in a weekly intoxication of Sabbatical sport, under the delirious form of the *chasse aux goujons*.

Clean as the town within the circuit of the river is, the houses which overhang the water on the other side are picturesque and dirty in the extreme, story rising above story, and balcony above balcony. It does not increase their beauty, and to a fastidious nose it must militate against their eligibility as places of residence, that there is apparently but one drain, an external one, which follows the course of the pillars supporting the various balconies: nevertheless, from the opposite side of the river, and when the wind sets the other way, they are sufficiently attractive. In this quarter is found the finest church, the Madeleine, with a very effective piece of sculpture at the east end. The sculpture is arranged on the bottom and farther side of a sort of cage, which is hung outside the church, but is visible from the inside through a corresponding opening in the east wall. The subject of the sculpture is 'The Sepulchre,' and the ends of the cage or box are composed of rich yellow glass, through which the external light streams into the cave of the Sepulchre; and when the church itself is becoming dark, the effect produced by the light from the evening sky, passing through the deep-toned glass, and softly illuminating the Sepulchre, is indescribably solemn.

When Besançon was supplied by the aqueduct with the waters of Arcier, there was a great abundance of baths, as the remains discovered in digging new foundations show; but in the present state of the town such things are not easily met with. The floating baths on the river are appropriated to the other sex, and the only thing approaching to a male

---

the Grand' Eau, with a torch in one hand to draw the fish to the top, and a sword in the other to kill them when they arrive there; a second man wading behind with a bag, to pick up the pieces.

[40] 'Swift-foot Almond, and land-louping Braan.'

bath was of a nature entirely new to me, being constructed as follows:-- There is a water-mill in the town, with a low weir stretching across the river, down which the water rushes with no very great violence. At the foot of this weir a row of sentry-boxes is placed, approached by planks, and in these boxes the adventurer finds his bath.[41]

Figure 9. Bath in the Doubs, at Besançon.

A stout piece of wood-work is fixed horizontally along the face of the weir, and has the effect of throwing the downward water out of its natural direction, and causing it to describe an arch, so that it descends with much force on to the weir at a point below the wood-work. Here two planks are placed, forming a seat and a support for the back, and a little lower still another plank for the feet to rest upon, without which the bather would have a good chance of being washed away. The water boils noisily and violently on all sides and in all directions, coming down upon the subject's shoulders with a heavy thud, which calls to mind the tender years when something softer than a cane was used, and sends him forth like a fresh-boiled lobster. All this, with towels, is not dear at fourpence.

The citadel is the great sight of Besançon, and the polite Colonel-commandant attends at his office at convenient hours to give passes. What it might be to storm the position under the excitement of the sport of war, I cannot say; but certainly it is a most trying affair on a hot Sunday's

[41] The sentry-box is omitted in the accompanying illustration.

afternoon, even when all is made smooth, and the gates are opened, by a comprehensive pass. The wall mentioned by Cæsar as a great feature of the place cut the site of the citadel off from the town, and many signs of it were found when the cathedral of S. Stephen was built, the unfortunate church which went down before the exigencies of a siege under Louis XIV. The barrack-master proved to be a most interesting man, knowing many details of Cæsar's life and campaigns which I suspect were not known to that captain himself. He had served in Algeria, and assented to the proposition that more soldiers died there of absinthe than of Arabs, stating his conviction that three-fourths of the whole deaths are caused by that pernicious extract of wormwood, and that he ought himself to have died of it long ago. He pointed out the difference between the massive masonry of the period of the Spanish occupation and the less impressive work of more recent times, and showed the dungeon from which Marshal Bourmont bought his escape, in the time of the first Napoleon.

The floor of one of the little look-out towers is composed of a tombstone, representing a priest in full ecclesiastical dress, and my question as to how it came there elicited the following story:--When Louis XIV. was besieging the citadel, he placed his head-quarters, and a strong battery, on the summit of the Mont Chaudane,[42] which commands the citadel on one side as the Brégille does on the other. Among the besieged was a monk named Schmidt, probably one of the Low-country men to whom the Franche Comté was then a sort of home, as forming part of the dominions of Spain; and this monk was the most active supporter of the defence, against the large party within the walls which was anxious to render the town. He was also an admirable shot; and on one of the last days of the siege, as he stood in the little tower where the tombstone now lies, the King and his staff rode to the front of the plateau on the Mont Chaudane to survey the citadel; whereupon someone pointed out to Schmidt that now he had a fair chance of putting an end at once to the siege and the invasion. Accordingly, he took a musket from a soldier and

---

[42] Believed to be derived from *Collis Dianæ*. Dunod found that *Chaudonne* was an early form of the name, and so preferred *Collis Dominarum*, with reference to the house of nuns placed there.

aimed at the King; but before firing he changed his aim, remarking, that he, a priest, ought not to destroy the life of a man, and so he only killed the horse, giving the Majesty of France a roll in the mud. When the town was taken, the King enquired for the man who killed his horse, and asked the priest whether he could have killed the rider instead, had he wished to do so. 'Certainly,' Schmidt replied, and related the facts of the case. Louis informed him, that had he been a soldier, he should have been decorated for his skill and his impulse of mercy; but, being a priest, he should be hung. The sentence was carried out, and the priest's body was buried in the floor of the tower from which he had spared the King's life. If this be true, it was one of the most unkingly deeds ever done.[43]

This siege took place in the second invasion or conquest of the Franche Comté by Louis XIV., when Besançon held out for nine days against Vauban and the King: on the first occasion it had surrendered to Condé after one day's siege, making the single stipulation that the Holy Shroud should not be removed from the town.[44] The *Saincte Suaire* was the richest ecclesiastical treasure of the Bisuntians, being one of the two most genuine of the many Suaires, the other being that of Turin, which was supported by Papal Infallibility. Both were brought from the Crusades; and the one was presented to Besançon in 1206, the other to Turin in 1353. Bede tells a story of the proving of a Shroud by fire in the eighth century, by one of the caliphs; and as its dimensions were 8 feet by 4, like that of Besançon, while the Shroud of Turin measured 12 feet by 3, the people of Besançon claimed that theirs was the one spoken of by Bede.

The Cathedral of Besançon is no longer S. Stephen, since the destruction of that church by Louis XIV. The small Church of the Citadel is now dedicated to that saint, an inscription on the wall stating that it takes

---

[43] Schmidt was not without the support of example in the indulgence of his warlike tastes. Thirty-eight years before, the religious took so active a part in the defence of Dôle against Louis XIII., that the Capuchin Father d'Iche had the direction of the artillery; and when an officer of the enemy had seized the Brother Claude by the cowl, the Father Barnabas made the officer loose his hold by slaying him with a demi-pique. When Arbois was besieged by Henry IV., the Sieur Chanoine Pécauld is specially mentioned as proving himself a *bon harquebouzier*.

[44] There is a painting by Vander Meulen, representing this siege, in the Fitzwilliam Museum in Cambridge.

the place of the larger church, *ex urbis obsidio anno 1674 lapsae*, and offering an indulgence of 100 days for every visit paid to it, with the sensible proviso *una duntaxat vice per diem.* Soldiers not being generally made of the confessing sex, or of confessing material, there is only one confessional provided for the 6,000 souls which the citadel can accommodate.

The Cavalry Barracks are in the lower part of the town, and near them is a large building with evident traces of ecclesiastical architecture on the outside. It is, in fact, a very fine church converted into stables, retaining its interior features in excellent preservation. Under the corn-bin lies a lady who had two husbands and fifteen children, *Antigone in parentes, Porcia in conjuges, Sempronia in liberos*; while a few yards further east, less agreeably placed, is an ecclesiastic of the Gorrevod family, who reckoned Prince and Bishop and Baron among his titles. The nave of this Church of S. Michael accommodates thirty horses, and the north aisle thirteen; the south is considered more select, and is boarded off for the decani, in the shape of officers' chargers. The north side of the chancel gives room for six horses, and the south side for a row of saddle-blocks. It had been an oversight on the part of the original architect of the church that no place was prepared for the daily hay; a fault which the military restorers have remedied by improvising a lady-chapel, where the hay for the day is placed in the morning. With Spelman in my mind, I asked if the stables were not unhealthy; but the soldiers said they were the healthiest in the town.[45]

The Glacière of Vaise had proved, as has been seen, to be a mare's-nest; and yet, after all, it produced a foal; for while I was endeavouring to overcome the evening heat of Besançon in a *spécialité* for ice, I found that the owner of the establishment was also the owner of the two glacières of Vaise; and in the course of the conversation which followed, he told me of the existence of a natural glacière near the village of Arc-sous-Cicon, twenty kilomètres from Pontarlier, which he had himself seen. As I had arranged to meet my sisters at Neufchâtel, in two days' time, for the

---

[45] The Church of S. Philibert, in Dijon, now a forage magazine, has an inscription let into the wall almost ludicrously out of keeping with the present desecrated state of the building,—*Dilexi Domine Decorem Domus tuæ*, 1648.

purpose of visiting a glacière in the Val de Travers, this piece of information came very opportunely, and I determined to attempt both glacières with them.

Some of the trains from Besançon stop for an hour at Dôle in passing towards Switzerland by way of Pontarlier, and anyone who is interested in the Burgundian and Spanish wars of France should take this opportunity of seeing what may be seen of the town of Dôle and its massive church-tower. The sieges of Dôle made it very famous in the later middle ages, more especially the long siege under Charles d'Amboise, at the crisis of which that general recommended his soldiers to leave a few of the people for seed,[46] and the old sobriquet *la Joyeuse* was punningly changed to *la Dolente*. It has had other claims upon fame; for if Besançon possessed one of the two most authentic Holy Shrouds, Dôle was the resting-place of one of the undoubted miraculous Hosts, which had withstood the flames in the Abbey of Faverney. It was for the reception of this Host that the advocates of the Brotherhood of Monseigneur Saint Yves built the Sainte Chapelle at Dôle.[47]

[46] 'Qu'on les laisse pour grain!'

[47] In the year 1648, it was suspected that some decay was going on in the material of this Host, and the following translation from the Latin describes the investigation entered into by the Dean and a large body of clergy and laity, in order to quiet the public mind:—'Après que tous les susnommés (viz. the Dean, Canons, President of the Parliament, &c.) étant présents eurent adorés le S. Sacrement, la custode fut ouverte avec tout le respect possible; et alors le dit Doyen aperçut un vermisseau roulé en spirale, qu'il saisit avec la pointe d'une épingle et plaça sur un corporal où chacun l'examina; puis on le brûla avec un charbon pris dans l'encensoir, et ses cendres furent jetées dans la piscine. On put alors constater tout le dommage que ce misérable petit animal avait causé aux espèces sacrées dont les débris ici tombaient en poussière, là se trouvaient rongés et lacérés, de telle sorte que l'Hostie n'avait presque plus rien de sa forme circulaire, et présentait de profondes découpures partout où le vermisseau s'était livré à ses sinueus es évolutions.'

Chapter 7

The Glacière of Monthézy, in the Val De Travers

I rejoined my sisters at Neufchâtel on the 5th of July, and proceeded thence with them by the line which passes through the Val de Travers. One of them had been at Fleurier, in 1860, on the day of the opening of this line, and she added an interest to the various tunnels, by telling us that a Swiss gentleman of her acquaintance, who had taken a place in one of the open carriages of the first train, found, on reaching the daylight after one of the tunnels, that his neighbour had been killed by a small stone which had fallen on to his head. Where the stone came from, no one could say, nor yet when it fell, for the unfortunate man had made no sign or movement of any kind.

Everyone must be delighted with the wonders of the line of rail, and the beauties through which the engineer has cut his way. In valleys on a less magnificent scale, cuttings and embankments on the face of the hill are sad eyesores, as in railway-ruined Killiecrankie; but here Nature's works are so very grand, that the works of man are not offensively prominent, being overawed by the very facts over which they have triumphed. When we reached the more even part of the valley, where the Reuse no longer roars and rushes far below, but winds quietly through the soft grass on a level with the rail, the whole grouping was so exceedingly charming, and the river itself so suggestive of lusty trout, and the village of Noiraigue[48] looked so tempting as it nestled in a sheltered nook among the headlong precipices, that I registered in a safe mental pigeon-hole a week at the auberge there with a fishing-rod, and excursions to the commanding summit in which the *Creux de Vent* is found. The engine-driver knew that he was in a region of beauties, and, when he whistled to warn his passengers that the train was about to move on, he remained stationary until the long-resounding echoes died out, floating lingeringly up the valley to neighbouring France.

We had no definite idea as to the *locale* of the glacière we were now bent upon attacking. M. Thury's list gave the following information:--'*Glacière de Motiers, Canton de Neufchâtel, entre les vallées de Travers et de la Brévine, près du sentier de la Brévine*;' and this I had rendered somewhat more precise by a cross-examination of the guard of the train on my way to Besançon. He had not heard of the glacière, but from what I told him he was inclined to think that Couvet would be the best station for our purpose, especially as the 'Ecu' at that place was, in his eyes, a commendable hostelry. Someone in Geneva, also, had believed that Couvet was as likely as anything else in the valley; so at Couvet we descended.[49]

This is a very clean and cheerful village, devoted to the lucrative manufacture of *absinthe*, and producing inhabitants who look like gentlemen and ladies, and promenade the ways in bonnets and hats, after a most un-Swiss-like fashion. They carefully restrict themselves to the making of the poisonous product of their village, and have nothing to do with the consumption thereof:[50] hence nature has a fair chance with them, and they are a healthy and energetic race. The beauties of the surrounding mountains, with their fitful alternations of pasture and wood, and grey face of rock, are not marred by the outward appearance, at least, of that which Bishop Heber lamented in a country where 'every prospect pleases.' An old

---

[48] *Aigue*, or *egue*, in the patois of this district, is equivalent to *eau*, the Latin *aqua*.

[49] Ebel, in his *Swiss Manual* (French translation of 1818, t. iii.), mentions this glacière under the head *Motiers*, and observes that it and the grotto of S. Georges are the only places in the Jura where ice remains through the summer. This statement, in common with a great part of Ebel, has been transferred to the letterpress of *Switzerland Illustrated*.

[50] Switzerland sent 7,500,000 gallons of absinthe to France in 1864.

lady is commemorated in the annals of Couvet as an example of the healthiness of the situation, who saw seven generations of her family, having known her great-grandfather in her early years, and living to nurse great-grandchildren in her old age. The landlord of the inn informed us, with much pride, that Couvet was the birthplace of the man who invented a clock for telling the time at sea; by which, no doubt, he meant the chronometer, invented by M. Berthoud. At Motiers, the next village, Rousseau wrote his *Lettres de la Montagne*, and thence it was that he fled from popular violence to the island on the Lake of Bienne.

The 'Ecu' promised us dinner in half an hour, and we strolled about in the garden of that unsophisticated hotel for an hour and a half, reconciled to the delay by the beauty of the neighbouring hills, the winding of the valley giving all the effect of a mountain-locked plain, with barriers decked with firs. It will readily be conceived, however, that three practical English people could not be satisfied to feed on beauty alone for any very great length of time, and we caught the landlady and became peremptory. She explained that dinner was quite ready, but she had intended to give us the pleasure of an agreeable society, consisting of sundry Swiss who were due in another half-hour or so: she yielded, nevertheless, to our representations, and promised to serve the meal at once. We were speedily summoned to the *salle-à-manger,* and entered a low smoke-stained wooden chamber, with no floor to speak of, and with huge beams supporting the roof, dangerous for tall heads. The date on the door was 1690, and the chamber fully looked its age. There was a long table of the prevailing hue, with a similar bench; and on the table three large basins, presumably containing soup, were ranged, each covered with its plate, and accompanied by a ricketty spoon of yellow metal and a hunch of black bread. A., who was hungry enough and experienced enough to have known better, began promptly a most pathetic 'Why surely!' but the landlady stopped her by opening a side door, and displaying a comfortable room in which a well-appointed table awaited us:--she had taken us through the kitchen rather than through the *salon*, in which were peasants smoking. We were somewhat disconcerted when we heard that the unwashed-looking place was the kitchen; but the landlady had made up for it by scrubbing her husband, who waited upon us, to a high pitch of presentability, and further experience showed that the 'Ecu' is to be highly commended for the excellence and abundance and cheapness of its foods.

There are many natural curiosities in and near the Val de Travers, which well repay the labour that must be expended upon them. The *Temple des Fées*, on the western side of the Valley of Verrières, used to be called the most beautiful grotto in Switzerland; and the great Cavern of La Baume, near Motiers, is said to be exceedingly wonderful. We were shown the entrance to a line of caverns in the hills above Couvet, and were informed that it was possible to pierce completely through the range, and pass out at the other side within sight of Yverdun. One of the caverns in this valley had been explored by some of A. and M.'s Swiss friends, and the account of what they had gone through was by no means inviting, seeing that the prevailing material was damp clay of a solid character, arranged in steep slopes, up which progression must be made by inserting the fingers and toes as far as might be into the clay; and, of course, when the handful of unpleasant mud came away, the result was the reverse of progression. To anyone who has only known the rope up the pure white side of some snow mountain, the idea of being roped for the purpose of grappling with underground banks of adhesive mud and clay must be horrible in the extreme. Another interesting natural phenomenon is presented by the source of the Reuse, that river gushing out from the rock in considerable volume, probably formed by the drainage of the Lake of Etallières, in the distant valley of La Brévine; while the Longe-aigue, on the contrary, is lost in a gulf of such horror that the people call the mill which stands on its edge the *Moulin d'enfer*.

As usual, we were assured that many of these remarkable sights were far better worth a visit than the glacière, of which no one seemed to know anything. A guide was at length secured for the next morning, who had made his way to the cave once in the winter-time and had been unable to enter it, and we settled down quietly to an evening of perfect rest. The windows of the bedrooms being guiltless of blinds and curtains, the effect of waking, in the early morning, to find them blocked up, as it were, by the green slopes of pasture and the dark bands of fir-woods which clothed the limiting hills, seemed almost magical, the foreground being occupied solely by the graceful curve of the dome of the church-tower, glittering with intercepted rays, and forming a bright omen for the day thus ushered in.

In due time the promised guide appeared, a sickly boy of unprepossessing appearance, and of *patois* to correspond. I was at first tempted to propose that we should attack him stereoscopically, A. administering French and I simultaneous German, in the hope that the combination might convey some meaning to him; but, after a time, we succeeded with French alone. Perhaps Latin would have made a more likely *mélange* than German, and to give it him in three dimensions would not have been a bad plan. The route for the glacière runs straight up the face of the hill along which the railway has been constructed; and as we passed through woods of beech and fir, with fresh green glades rolling down below our feet, or emerged from the woods to cross large undulating expanses of meadow-land, we were almost inclined to believe that we had never done so lovely a walk. The scenery through which we passed was thoroughly that of the lower districts of the Alps, with nothing Jurane in its character, and the elevation finally achieved was not very great: indeed, at a short distance from the glacière, we passed a collection of very neat châlets, with gardens and garden-flowers, one of the châlets rejoicing in countless beehives, with three or four 'ekes' apiece. Up to the time of reaching this little village, which seemed to be called Sagnette, our path had been that which leads to *La Brévine*, the highest valley in the canton; but now we turned off abruptly up the steeper face on the left hand, and in a very few minutes came upon a dry wilderness of rock and grass, which we at once recognised as 'glacière country;' and when I told our guide that we must be near the place, he replied by pointing to the trees round the mouth of the pit.

Shortly after we first left Couvet, a gaunt elderly female, with a one-bullock char, had joined our party, and tried to bully us into giving up the cave and going instead to a neighbouring summit, whence she promised us a view of unrivalled extent and beauty. She told us that there was nothing to be seen in the glacière, and that it was a place where people lost their lives. The guide said that was nonsense; but she reduced him to silence by quoting a case in point. She said, too, that if a man slipped and fell, there was nothing to prevent him from going helplessly down a run of ice into a subterranean watercourse, which would carry him for two or three leagues underground; and on this head our boy had no counter-statement to make. She asserted that without ladders it was utterly impossible to make the descent to the commencement of the glacière; and she vowed there was no ladder now, nor had been for some time. Here the boy came in, stating that the cave belonged to a mademoiselle of Neufchâtel, who had a summer cottage at no great distance, and loved to be supplied with ice during her residence in the country, for which purpose she kept a sound ladder on the spot, and had it removed in the winter that it might not be destroyed. There was a circumstantial air about this statement which for the moment got the better of the old woman; but she speedily recovered herself, and repeated positively that there was no ladder of any description, adding, somewhat inconsequently, that it was such a bad one, no Christian could use it with safety. The boy retorted, that it was all very well for her to run the glacière down, as she lived near it, but for the world from a distance it was a most wonderful sight; and, as for the ladder, he happened to know that it was at this time in excellent preservation. The event proved that in saying this he drew entirely upon his imagination. It is, perhaps, only fair to suppose that they don't mean anything by it, and it may be mere ignorance on their part; but the simple fact is, that some of those Swiss rustics tell the most barefaced lies conceivable,--*unblushing* is an epithet that cannot be safely applied without previous soap and water,--and tell them in a plodding systematic manner which takes in all but the experienced and wary traveller. I have myself learned to suspend my judgment regarding the most simple thing in nature, until I have other grounds for forming an opinion than the solemn asseverations of the most stolid and respectable Swiss, if it so be that money depends upon his report.[51]

---

[51] *Point d'argent, point de Suisse*, is a proverbial expression which the Swiss twist into a historical compliment, asserting that it arose in early mercenary times, from the fact that they were too virtuous to accept the suggestion of the general who hired them, and wished them to take their pay in kind from the defenceless people of the country they had served.

As in the case of two of the glacières already described, the entrance is by a deep pit, which has the appearance of having been at one time two pits, one less deep than the other; and the barrier between the two having been removed by some natural process, a passage is found down the steep side of the shallower pit, which lands the adventurer on a small sloping shelf, 21 feet sheer above the surface of the snow in the deeper pit, the sides of the latter rising up perpendicularly all round. It is for this last 21 feet that some sort of ladder is absolutely necessary. Our guide flung himself down in the sun at the outer edge of the pit, and informed us that as it was cold and dangerous down below, he intended to go no farther: he had engaged, he said, to guide us to the glacière, and he felt in no way bound to go into it. He was not good for much, so I was not sorry to hear of his determination; and when my sisters saw the sort of place they had to try to scramble down, they appeared to be very glad that only I was to be with them.

Leaving them to make such arrangements with regard to dress as might seem necessary to them, I proceeded to pioneer the way down the first part of the descent. This was extremely unpleasant, for the rocks were steep and very moist, with treacherous little collections of disintegrated material on every small ledge where the foot might otherwise have found a hold. These had to be cleared away before it could be safe for them to descend, and in other places the broken rock had to be picked out to form foot-holes; while, lower down, where the final shelf was reached, the abrupt slope of mud which ended in the sheer fall required considerable reduction, being far too beguiling in its original form. Here there was also a buttress of damp earth to be got round, and it was necessary to cut out deep holes for the hands and feet before even a man could venture upon the attempt with any comfort. The buttress was not, however, without its advantage, for on it, overhanging the snow of the lower pit, was a beautiful clump of cowslips (*Primula elatior*, Fr. *Primevère inodore*), which was at once secured as a trophy. The length of the irregular descent to this point was between 70 and 80 feet. On rounding the buttress, the upper end of the ladder presented [illegible] the question, between the boy and the old woman was to be [illegible]ked down to the edge of the shelf, and looked over into the pit, and, alas! the state of the remaining parts of the ladder was hopeless, owing partly to the decay of the sidepieces, and partly to the general absence of steps--a somewhat embarrassing feature under the circumstances. A further investigation showed that for the 21 feet of ladder there were only seven steps, and these seven were not arranged as conveniently as they might have been, for two occurred at the very top, and the other five in a group at the bottom. A branchless fir-tree had at some time fallen into the pit, and now lay in partial contact with the ruined ladder; and there were on the trunk various little knobs, which might possibly be of some use as a supplement to the rare steps of the ladder. The snow at the bottom of the pit was surrounded on all sides by perpendicular rock, and on the side opposite to the ladder I saw an arch at the foot of the rock, apparently 2 or 3 feet high, leading from the snow into darkness; and that, of course, was the entrance to the glacière. I succeeded in getting down the ladder, by help of the supplement, and looked down into the dark hole to see that it was practicable, and then returned to report progress in the upper regions. We had brought no alpenstocks to Couvet, so we sent the guide off into the woods, where we had heard the sound of an axe, to get three stout sticks from the woodmen; but he returned with such wretched, crooked little things, that A. went off herself to forage, and, having found an impromptu cattle-fence, came back with weapons resembling bulbous hedge-stakes, which she skinned and generally modified with a powerful clasp-knife, her constant companion. She then cut up the crooked sticks into *bâtons* for a contemplated repair of the ladder, while M. and I investigated the country near the pit. We found two other pits, which afterwards proved to communicate with the glacière. We could approach sufficiently near to one of these to see down to the bottom, where there was a considerable collection of snow: this pit was completely sheltered from the sun by trees, and was 66 feet deep and 4 or 5 feet in diameter. The other was of larger size, but its edge was so treacherous that we did not venture so near as to see what it contained: its depth was about 70 feet, and the stone and a foot or two of the string came up wet. The sides of the main pit, by which we were to enter the glacière, were, as has been said, very sheer, and on one side we could approach sufficiently ne

to the ancle' was its prevailing condition. The boy told us, when we rejoined him, that when he and others had attempted to get ice for the landlord, when it was ordered for him in a serious illness the winter before, they had found the pit filled to the top with snow.

As we stood at the mouth of the low entrance, making final preparations for a plunge into the darkness, I perceived a strong cold current blowing out from the cave--sufficiently strong and cold to render knickerbocker stockings a very unavailing protection. While engaged in the discovery that this style of dress is not without its drawbacks, I found, to my surprise, that the direction of the current suddenly changed, and the cold blast which had before blown out of the cave, now blew almost as strongly in. The arch of entrance was so low, that the top was about on a level with my waist; so that our faces and the upper parts of our bodies were not exposed to the current, and the strangeness of the effect was thus considerably increased.

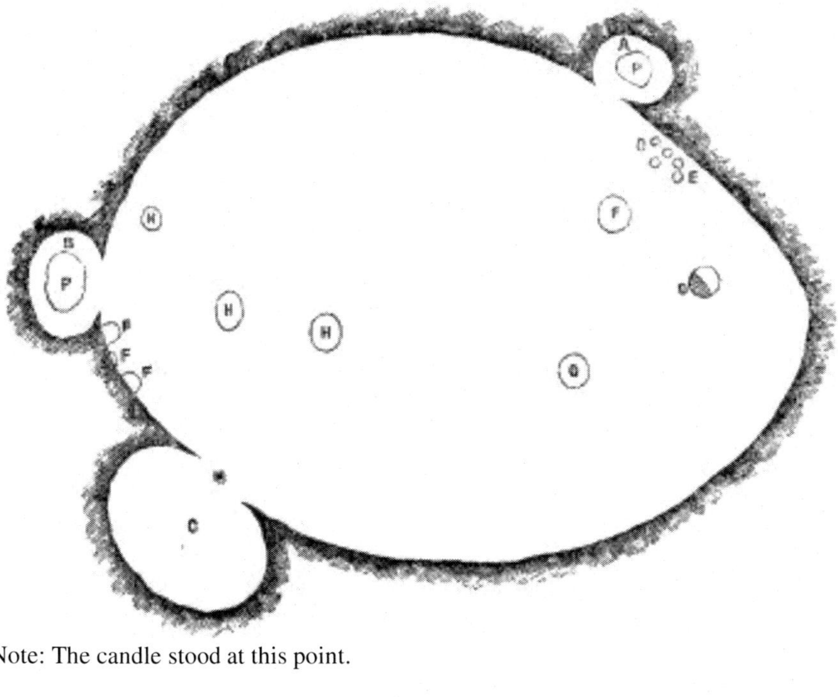

Note: The candle stood at this point.

Figure 11. Ground Plan of the Glacière of Monthézy.

As a matter of curiosity, we lighted a *bougie*, and placed it on the edge of the snow, at the top of the slope of 3 or 4 feet which led down the surface of the ice, and then stood to watch the effect of the current on the flame. The experiment proved that the currents alternated, and, as I fancied, regularly; and in order to determine, if possible, the law of this alternation, I observed with my watch the exact duration of each current. For twenty-two seconds the flame of the *bougie* was blown away from the entrance, so strongly as to assume a horizontal position, and almost to leave the wick: then the current ceased, and the flame rose with a stately air to a vertical position, moving down again steadily till it became once more horizontal, but now pointing in towards the cave. This change occupied in all four seconds; and the current inwards lasted--like the outward current--twenty-two seconds, and then the whole phenomenon was repeated. The currents kept such good time, that when I stood beyond their reach, and turned my back, I was enabled to announce each change with perfect precision. On one occasion, the flame performed its semicircle in a horizontal instead of a vertical plane, moving round the wick in the shape of a pea-flower. The day was very still, so that no external winds could have anything to do with this singular alternation; and, indeed, the pit was so completely sheltered by its shape, that a storm might have raged outside without producing any perceptible effect below. It would be difficult to explain the regularity of these opposite currents, but it is not so difficult to see that some such oscillation might be expected. It will be better, however, to defer any suggestions on this point till the glacière has been more fully described.

We passed down at length through the low archway, and stood on the floor of ice. As our eyes became accustomed to the darkness, we saw that an indistinct light streamed into the cave from some low point at a considerable distance, apparently on a level with the floor; and this we afterwards found to be the bottom of the larger of the two pits we had already fathomed, the pit A of the diagram; and we eventually discovered a similar but much smaller communication with the bottom of the pit B. In each of these pits there was a considerable pyramid of snow, whose base was on a level with the floor of the glacière: the connecting archway in the case of the pit A was 3 or 4 feet high, allowing us to pass into the pit and round the pyramid with perfect ease, while that leading to the pit B was less than a foot high, so that no passage could be forced.

As we stood on the ice at the entrance and peered into the comparative darkness, we saw by degrees that the glacière consisted of a continuous sea of smooth ice, sloping down very gently towards the right hand. The rock which forms the roof of the cave seemed to be almost as even as the floor, and was from 4 to 5 feet high in the neighbourhood in which we now found ourselves, gradually approaching the floor towards the bottom of the pit B, where it became about a foot high, and rising slightly in that part of the cave where the floor fell, so as to give 9 or 10 feet as the height there. The ice had all the appearance of great depth; but there were no means of forming a trustworthy opinion on this point, beyond the fact that I succeeded in lowering a stone to a considerable depth, in the small crevice which existed between the wall and the block of ice which formed the floor. The greatest length of the cave we found to be 112 ft. 7 in., and its breadth 94 ft., the general shape of the field of ice, which filled it to its utmost edges, being elliptical. The surface was unpleasantly wet, chiefly in the line of the currents, which were now seen to pass backwards and forwards between the pits A and C. In the neighbourhood of the pit B the water stood in a very thin sheet on the ice, which there was level, and rendered the style of locomotion necessitated by the near approach of the roof extremely disagreeable, as I was obliged to lie on my face, and push myself along the wet and slippery ice, to explore that corner of the cave, being at length stopped by want of sufficient height for even that method of progression.

The circle marked D represents a column from the roof, at the foot of which we found a small grotto in the ice, which I entered to a depth of 6 feet, the surface of the field of ice showing a very gracefully rounded fall at the edges of the grotto. At the point E there was a beautiful collection of fretted columns, white and hard as porcelain, arranged in a semicircle, with the diameter facing the cave, measuring 22 ft. 9 in. along this face. On the farther side of these columns there were signs of a considerable fall in the ice; and by making use of the roots of small stalagmitic columns of that material, which grew on the slope of ice, I got down into a little wilderness of spires and flutings, and found a small cave penetrating a short way under the solid ice-floor. G marks the place of a free stalagmite of ice, formed under a fissure in the roof; and each F represents a column from the roof, or from a lateral fissure in the wall.

The most striking features of this cave were the three domes, marked H in the ground-plan, in which they ought strictly not to appear, as being confined to the roof: one of them is shown also in the vertical section of the cave. They occur where the roof is from 3 to 4 feet above the floor. It will be understood, that the bent attitude in which we were obliged to investigate these parts of the cave was exceedingly fatiguing, and we hailed with delight a sudden circular opening in the roof which enabled us to stand upright. This delight was immensely increased when our candles showed us that the walls of this vertical opening were profusely decorated with the most lovely forms of ice. The first that we came under passed up out of sight; and in this, two solid cascades of ice hung down, high overhead, apparently broken off short, or at any rate ending very abruptly: the others did not pass so far into the roof, and formed domes of very regular shape. In all three, the details of the ice-decoration were most lovely, and the effect produced by the whole situation was very curious; for we stood with our legs exposed to the alternating cold currents, the remaining part of our bodies being imbedded as it were in the roof; while the candles in our hands brought out the crystal ornaments of the sides, flashing fitfully all round us and overhead, when one or other of us moved a light, as if we had been surrounded by diamonds of every possible size and setting. One of the domes was so small, that we were obliged to stand up by turn to examine its beauties; but in the others we all stood together. On every side were branching clusters of ice in the form of club-mosses, with here and there varicose veins of clear ice, and pinnacles of the prismatic structure, with limpid crockets and finials. The pipes of ice which formed a network on the walls were in some cases so exquisitely clear, that we could not be sure of their existence without touching them; and in other cases a sheet 4 or 6 inches thick was found to be no obstruction to our view of the rock on which it was formed. In one of the domes we had only one candle, and the bearer of this after a time contrived to let it fall, leaving us standing with our heads in perfect darkness; while the indistinct light which strayed about our feet showed faintly a circle of icicles, hanging from the lower part of the dome, the fringe, as it were, of our rocky petticoats.

In one of the lower parts of the cave, where darkness prevailed, and locomotion was only possible on the lowest reptile principles, M. announced that she could see clear through the ice-floor, as if there were nothing between her and the rock below. I ventured to doubt this, for there was an air of immense thickness about the whole ice; and as soon as A. and I had succeeded in grovelling across the intervening space, and converged upon her, we found that the appearance she had observed was due to a most perfect reflection of the roof, as shown by the candles we carried, which may give some idea of the character of the ice. We did not care to study this effect for any very prolonged time, inasmuch as we were obliged meanwhile to stow away the length of our legs on a part of the ice which was thinly covered with water,--one result of its proximity to the arch communicating with the smallest pit.

It has been said that the whole ice-floor sloped slightly towards one side of the cave, the slope becoming rather more steep near the edge.[52] Clearly, ever so slight a slope would be sufficiently embarrassing, when the surface was so perfectly smooth and slippery; and this added much to the difficulty of walking in a bent attitude. On coming out of one of the domes, I tried progression on all-fours--threes, rather, for the candle occupied one hand,--and I cannot recommend that method, owing to the impossibility of putting on the break. The pace ultimately acquired is greater than is pleasant, and the roof is too near the floor to allow of any successful attempt to bring things to an end by the reassumption of a biped character.

We placed a thermometer in the line of greatest current, and another in a still part of the cave. The memorandum is lost of their register--if, indeed,

[52] It is probable that the ice is on the increase in this glacière, and that an archway, now filled up by the growing ice, has at one time existed in the wall on this side of the care, through which the ice and water used to pour into the subterranean depths of which the old woman had told us. At the time of our visit, we could find no outlet.

we ever made one, for we were more concerned with the beauties than the temperature was surprisingly high in the line of current, as compared with the ordinary temperature of ice-caves.

When we came to compare backs, after leaving the cave, we mutually found that they were in a very disreputable condition. The damp and ragged roof with which they had been so frequently in contact had produced a marked effect upon them, and I eventually paid a tailor in Geneva three francs for restoring my coat to decency. M. took great credit to herself for having been more careful of her back than the others, and declined to be laughed at for forgetting that she was only about half as high as they, to begin with. A. still remembers the green-grey stains, as the most obstinate she ever had to deal with, especially as her three-days' knapsack contained no change for that outer part of her dress.

The 'Ecu' gave us a charming dinner on our return; then a moderate bill, and an affectionate farewell; and we succeeded in catching the early evening train for Pontarlier.[53]

---

[53] The following remarks may give some explanation of the phenomenon of alternating currents in this cave, I should suppose that during the night there is atmospheric equilibrium in the cave itself, and in the three pits A, B, C. When the heat of the sun comes into operation, the three pits are very differently affected by it, C being comparatively open to the sun's rays, while A is much less so, and B is entirely sheltered from radiation. This leads naturally to atmospheric disturbance. The air in the pit C is made warmer and less heavy than that in A and B, and the consequence is, that the column of air in C can no longer balance the columns in A and B, which therefore begin to descend, and so a current of air is driven from the cave into the pit C. Owing to the elasticity of the atmosphere, even at a low temperature, this descent, and the consequent rush of air into C, will be overdone, and a recoil must take place, which accounts for the return current into the cave from the pit C. The sun can reach A more easily than B, and thus the air is lighter and more moveable in the former pit, so that the recoil will make itself more felt in A than in B: accordingly, we found that the main currents alternated between A and C, with very slight disturbance in the neighbourhood of B. B will, however, play its part, and the weighty column of air contained in it will oscillate, though with smaller oscillations than in the case of A. Probably, when the sun has left A, while acting still upon C, the return current from C will be much slighter, and there will be a general settling of the atmosphere in the pits A and B, until C also is freed from the sun's action, when the whole system will gradually pass into a state of equilibrium.

With respect to the action of the more protected pits, the principle of the hydraulic ram not unnaturally suggests itself. In considering the minor details of the currents, such elements as the refrigeration of the air in its passage across the face of the ice must be taken into account. It may be observed that the candle did not occupy an *intermediate* position with respect to two opposing currents, for it was practically on the floor of the cave, owing to the continuity of the slope of snow on which it stood, as shown in the vertical section on p. 108.

Chapter 8

THE GLACIÈRE AND NEIGIÈRE OF ARC-SOUS-CICON

The beauties of the Val de Travers end only with the valley itself, at the head of which a long tunnel ushers the traveller into a tamer country,--a preparation, as it were, for France. After the border is passed, the scenery begins to improve again, and the effect of the two castles of Joux, the new and the old, crowning the heights on either side of the narrow gorge through which the railway runs, is very fine. The guide-books inform us that the Château of Joux was the place of imprisonment of the unfortunate Toussaint L'Ouverture, and that there he died of neglect and cold; and it was in the same strong fortress that Mirabeau was confined by his father's desire. The old castle, however, is more interesting from its connection with the history of Charles the Bold, who retired to La Rivière after the battle of Morat, and spent here those sad solitary weeks of which Philip de Comines tells with so many moral reflections; weeks of bodily and mental distress, which left him a mere wreck, and led to his wild want of generalship and his miserable death at Nancy. He had melted down the church-bells in this part of Burgundy and Vaud, to make cannon for the final effort which failed so fatally at Morat; and the old chroniclers relate--without any allusion to the sacrilege--that the artillery was wretchedly served on that cruel[54] day. It is some comfort to Englishmen to know that their ancestors under the Duke of Somerset displayed a marvellous courage on the occasion.

We reached Pontarlier in time for a stroll through the quiet town; but we searched in vain for the tempting convents and gates, which were marked on my copy of an old plan of the place, dedicated to the Prince d'Arenberg, in the well-known times when he governed the Franche Comté. The convents had become for the most part breweries, and the gates had been improved away. Our enquiries respecting the place of our destination were fortunately more successful. The idea of a glacière was new to the world of Pontarlier; but the landlord of the Hôtel National had heard of Arc-sous-Cicon, and had no doubt that we could find a carriage of some sort to take us there. His own horses were all engaged in haymaking, but his neighbours' horses might be less busy, and accordingly he took us first to call upon M. Paget, a friend who added to his income by keeping a horse and voiture for hire. The Pagets in general had gone to bed, and the door was fastened; but our guide seemed to know the ways of the house, and we found Madame in the stables, and arranged with her for a carriage at seven o'clock the next morning.

At the time appointed, M. Paget did not come, and I was obliged to go and look him up. He proved to me that it was all right, somehow, and evidently understood that his convenience, not ours, was the thing to be consulted. The hotel is in a narrow street, and, apparently on that account, a stray passer-by was caught, and pressed into M. Paget's service to help to turn the carriage,--a feat accomplished by a bodily lifting of the hinder part, with its wheels. After-experience showed that the narrowness of the street had nothing to with it, and we discovered that the necessity for the manoeuvre was due to a chronic affection of some portion of the voiture; so that whenever in the course of the day it became necessary for us to turn round, M. Paget was constrained to call in foreign help.

The country through which we passed was uninteresting in the extreme, although we had been told by the landlord that our drive would

---

[54] *Cruel comme à Morat* was long a popular saying.

introduce us to a succession of natural beauties such as few countries in the world could show. The line of hills, at the foot of which we expected our route to lie, looked exceedingly tempting as seen from Pontarlier; but, to our disappointment, we left the hills and struck across the plain. About ten or eleven kilomètres from Pontarlier, however, the character of the country changed suddenly, and we found the landlord's promise in some part fulfilled. Rich meadow-slopes were broken by solitary trees arranged in Nature's happiest style, and grey precipices of Jurane grimness and perpendicularity encroached upon the woods and grass. We were coming near the source of the Loue, M. Paget said, which it would be necessary for us to visit. He told us that we must leave the carriage at an *auberge* on the roadside, and walk to the neighbouring village of Ouhans, which was inaccessible for voitures, and thence we should easily find our way to the source. The distance, he declared, was twenty minutes. The woman at the *auberge* strongly recommended the source, but did her best to dissuade us from the glacières, of which she said there were two. She had visited them herself, and told her husband, who had guided her, that there was nothing to see. That, we thought, proved nothing against the glacières, and her dulness of appreciation we were willing to accept without further proof than her personal appearance. Besides, to go to the source, and not to Arc, would mean dining with her; so that she was not an impartial adviser.

M. Paget was a short square man, of very few words, and his one object in life seemed to be to save his black horse as much as possible; a very creditable object in itself, so long as he did not go too far in his endeavours to accomplish it. On the present occasion he certainly did go too far. The road was quite as good as that which we had left, and there was no reason in the world why the carriage should not have taken us to the village. Worse still, we discovered eventually that the 'twenty minutes' meant twenty minutes from the village to the source, and represented really something like half the time necessary for that part of the march, while there was a hot and dusty walk of half an hour before we reached the village. As he accompanied us in person, we had the satisfaction of frequently telling him our mind with insular frankness. He pretended to be much distressed, but assured us each time we returned to the charge--about every quarter of an hour--that we were close to the desired spot. From the village to the source, the way led us through such pleasant scenery and such acceptable strawberries, that we only kept up our periodical remonstrances on principle, and, after we had wound rapidly down through a grand defile, and turned a sudden angle of the rock, the first sight of that which we had come to see amply repaid us all the trouble we had gone through. The source of the Orbe is sufficiently striking, but the Loue is by far more grand at the moment of its birth. The former is a bright fairy-like stream, gushing out of a small cavern at the foot of a lofty precipice clothed with clinging trees; but the Loue flows out from the bottom of an amphitheatrical rock much more lofty and unbroken. The stream itself is broader and deeper, and glides with an infinitely more majestic calmness from a vast archway in the rock, into the recesses of which the eye can penetrate to the point where the roof closes in upon the water, and so cuts off all further view. The calmness of the flow may be in part attributed to a weir, which has been built across the stream at the mouth of the cave, for the purpose of driving a portion of the water into a channel which conveys it to various mill-wheels; for, at a very short distance below the weir, the natural stream makes a fall of 17 feet, so that, if left to itself, it might probably rush out more impetuously from its mysterious cavern. The weir is a single timber, below the surface, fixed obliquely across the stream on a shelving bank of masonry, and the farther end meets the wall of rock inside the cave. Near it we saw some glorious hart's-tongue ferns, which excited our desires, and I took off boots and stockings, and endeavoured to make my way along the weir; but the face of the masonry was so very slippery, and the nails in the timber so unpleasant for bare feet, and the stream was so unexpectedly strong, that I called to mind the proverbial definition of the better part of valour, and came back without having achieved the ferns. The biting coldness of the water, and the boiling of the fall close below the weir, did not add to my confidence in making the attempt, but I should think that in a more favourable state of the water the cave might be very well explored by two men going alone. The day penetrated so completely into the farthest corners, that when I got half-way along the weir, I could detect the oily look on the surface where it first saw the light, which showed where the water was quietly streaming up from its unknown sources. The people in the neighbourhood were unable to suggest any lake or lakes of which this river might be the subterranean drainage. It is liable to sudden and violent overflows, which seldom last more than twenty-four hours; and from the destruction of property caused by these outbursts, the name of *La Loue*, sc. *La Louve*, has been given to it. The rocky valley through which the river runs, after leaving its underground channel, is exceedingly fine, and we wandered along the precipices on one side, enjoying the varying scenes so much that we could scarcely bring ourselves to turn; each bend of the fretting river showing a narrow gorge in the rock, with a black rapid, and a foaming fall. It is said that although the mills on the Doubs are sometimes stopped from want of water, those which derive their motive power from this strange and impressive cavern have never known the supply to fail.

Before we started for our ramble among the woods and precipices which overhang the farther course of the Loue, we had sent off M. Paget to the *auberge*, with strict orders that he should at once get out the black horse, and bring the carriage to meet us at Ouhans, as one of us was not in so good order for walking as usual, and the day was fast slipping away. Of course we saw nothing of him when we reached Ouhans; and as it was not prudent to wait for his arrival there, which might never take place, we walked through the broiling sun in the direction of the *auberge,* and at last saw him coming, pretending to whip his horse as if he were in earnest about the pace. We somewhat sullenly assisted him to turn the old carriage round, and then bade him drive as hard as he could to Arc-sous-Cicon, still a long way off. This he said he would do if he knew which was the way; but since he was last there, as a much younger man, there had been a general change in the matter of roads, and how the new ones lay he did not know. This was not cheerful intelligence, especially as we had set our hearts upon getting back to Pontarlier in time for the evening train, which would give us a night at the charming *Bellevue* at Neufchâtel, instead of the poisonous coffee and the trying odours of the *National*: the old man's instinct, however, led him right, and we reached Arc at half-past twelve. One obstacle to our journey on the new road promised at first to be insurmountable, being an immense *sapin*, the largest I have seen felled, which lay on a combination of wood-chairs straight across the road. It had been brought down a narrow side-road through a wheat-field, and one end occupied this road, while the other was jammed against the wall on the opposite side of the main road; and half-a-dozen men, with as many draught oxen, were mainly endeavouring to turn it in the right direction. M. Paget knew how much was required to turn his own carriage, and he calculated that the road would not be free for two or three hours, which involved a rest for his black horse, a pipe for himself, and, possibly, a short sleep. The oxen were lazy, and their hides impervious; the whips were cracked in vain, and in vain were brought more directly to bear upon the senses of the recusants; the men howled, and rattled the chains, and re-arranged the clumsy head-gear, but all to no purpose. The man who did most of the howling was a black Burgundian dwarf, in a long blouse and moustaches; and he did it in so frightful a patois, that the oxen were right in their refusal to understand. We represented to M. Paget that it would be possible to make our way through the wheat; but he declared himself perfectly happy where he was, and declined to take any steps in the matter; whereupon I assumed the command of the expedition, and led the horse through the corn, thus turning the flank of the *sapin* and its attendants. Our driver submitted to this act of violence much as a member of the Society of Friends allows a chamberlain to remove his hat from behind when he is favoured with an audience of the sovereign; and when we regained the high road, he meekly took up the reins and drove us at a good pace to Arc.

The village lies in a curiously open plain, with a girdle of hills, in one of which the glacières were supposed to lie. The first *auberge* refused us admittance, on the ground that the dinner was all pre-engaged, and the result was that we found a pleasanter place higher up the village, near a vast new *maison de ville* with every window shattered by recent hail. The people groaned over the unnecessary expense of this huge building, which might well, from its size, have been a home for the whole village; and they told us that the communal forests had been terribly over-cut to provide the money for it. Our first demand was for food; our next, for a guide to the glacières. Food we could have; but why *should* we wish to go to the glacières, when there was so much else worth seeing at a little distance?--a guide might without doubt be found, but there was nothing to be seen when we got there. We ordered prompt dinner, anything that happened to be ready, and desired the landlord to look out for a man to show us the way up the hills. When the dinner came, it was cold; and the main dish consisted apparently of something which had made stock for many generations of soup, and had then been kept in a half-warm state, ready to be heated for any passer-by who called for hot meat, till the cook had despaired of its ever being used, and had allowed it to become cold: at least, no other supposition seemed to account for its utter want of flavour, and the wonderful development of its fibres. As a matter of politeness, I asked the man what it was; when he took the dish from the table, smelled at it, and pronounced it veal.

There were also several specimens of the original old turnip-radish, with large shrubs of heads, and mature feelers many inches long. As all this was not very inviting, we ordered an omelette and some cheese; and when the omelette came, we found that the cook had combined our ideas and understood our order to mean a cheese-omelette, which was not so bad after all.

By this time, the landlord's visit to his drinking-room had procured a man willing to act as our guide. He was, unfortunately, more willing than able; for his sojourn in the drinking-room had told upon his powers of equilibrium. He asserted, as everyone seemed in all cases to assert, that neither rope nor axe was in any way necessary. When I pressed the rope, he said that if monsieur was afraid he had better not go; so we told the landlord privately that the man was rather too drunk for a guide, and we must have another. The landlord thereupon offered himself, at the suggestion of his wife, who seemed to be the chief partner in the firm, and we were glad to accept his offer; while the incapacitated man whom we had rejected acquiesced in the new arrangement with a bow so little withering, and with such genuine politeness, that, in spite of his over-much wine, he won my heart. The landlord himself did not profess to know the glacières; but he knew the man who lived nearest to them, and proposed to lead us to his friend's châlet, whence we should doubtless be able to find a guide.

We stole a few moments for an inspection of the Church of Arc, and found, to our surprise, some very pleasing paintings in good repair, and open sittings which looked unusually clean and neat. Then we crossed the plain towards the north, and proceeded to grapple with a stiff path through the woods which climb the first hills. It turned out that there was no one available for our purpose in the châlet to which the landlord led us; but a small child was despatched in search of the master or the domestic, and returned before long with the latter individual, who received the mistress's instruction respecting the route, and received also an axe which I had begged in case of need. The accounts we had heard of the glacière or glacières--every one declined to call them caves--were so various, and the total denials of their existence so many, that we quietly made up our minds to disappointment, and agreed that what we had seen at the source of the Loue was quite sufficient to repay us for the trouble we had taken; while the idea of a rapid raid into France had something attractive in it, which more than counterbalanced the old charms of Soleure. Besides, we found that we were now in a good district for flowers, and the abundant *Gnaphalium sylvaticum* brought back to our minds many a delightful scramble in glacier regions, where its lovely velvet kinsman the *pied-de-lion* grows. On the broad top of the range of hills, covered with rich grass, we came upon large patches of a plant, with scented leaves and pungent seeds, which we had not known before, *Meum athamanticum*, and, to please our guide, we went through the form of pretending that we rather liked its taste. My sisters were in ecstasies of triumph over a wild everlasting-pea, which grew here to a considerable height--*Lathyrus sylvestris*, they said, Fr. *Gesse sauvage*, distinct from *G. hétéropyhlle,* which is still larger, and is almost confined to a favourite place of sojourn with us, the little Swiss valley of Les Plans. It is said that on the top of these hills springs of water rise to the surface, though there is no higher ground in the neighbourhood; a phenomenon which has been accounted for by the supposition of a difference of specific gravity between these springs and the waters which drive them up.

The character of the ground on the plateau changed suddenly, and we passed at one step, apparently, from a meadow of flowers to a wilderness of fissured rock, lying white and skeleton-like in the afternoon sun. We only skirted this rock in the first instance, and made for a clump of trees some little way off, in which we found a deep pit, with a path of sufficient steepness leading to the bottom. Here we came to a collection of snow, much sheltered by overhanging rocks and trees; and this, our guide told us, was the *neigière*, a word evidently formed on the same principle as *glacière*. The snow was half-covered with leaves, and was unpleasantly wet to our feet, so that we did not spend much time on it, or rather in it. A huge fragment of rock had at some time or other fallen from overhead, and now occupied a large part of the sloping bottom of the pit: by squeezing myself through a narrow crevice between this and the live rock, which looked as if it ought to lead to something, I found a veritable ice-cave, unhappily free from ornament, and of very small size, like a round soldier's tent in shape, with walls of rock and floor of ice. We afterwards found an easier entrance to the cave; but the floor was so wet, and the constant drops of water from the roof so little agreeable, that we got out again as soon as possible, especially as this was not the glacière we had come to see.

When we reached the surface once more, the landlord and the domestic both assured us that the *neigière* was the great sight, the glacière being nothing at all, but, such as it was, they would lead us to it. They took us to the fissured rock mentioned above; and when we looked down into the fissures, we saw that some of them were filled at the bottom with ice. They were not the ordinary fissures, like the crevasses of a glacier, but rather disconnected slits in the surface, opening into larger chambers in the heart of the rock, where the ice lay. In one part of this curious district the surface sank considerably, and showed nothing but a tumbled collection of large stones and rocks, piled in a most disorderly manner. By examining the neighbourhood of the larger of these rocks, we found a burrow, down which one of the men and I made our way, and thus, after some windings in the interior, reached a point from which we could descend to the ice. The impression conveyed to my mind by the whole appearance of the rock and ice was not unlike that of the domes in the Glacière of Monthézy; only that now the lower part of the dome was filled with ice, and we stood in the upper part. There were two or three of these domes, communicating one with another, and in all I found abundant signs of the prismatic structure, though no columns or wall-decoration remained. My sisters were accomplished in the art of burrowing, but they did not care to come down, and we soon rejoined them, spending a little time in letting down lighted *bougies* into the various domes and fissures, in order to study the movements of the air, but our experiments did not lead to much.

The landlord had evidently not believed in the existence of ice in summer, and his first thought was to take some home to his wife, to prove that we had reached the glacière and had found ice: such at least were the reasons he gave, but evidently his soul was imbued with a deep obedience to that better half, and the offering of a block of ice was suggested by a complication of feelings. When we reached the *auberge* again, we found the rejected guide still there, and more unstable than before. The general impression on his mind seemed to be that he had been wronged, and had forgiven us. In our absence he had been meditating upon the glacière, and his imagination had brought him to a very exalted idea of its wonders. Whereas, in the former part of the day, he had stoutly asserted that no cord could possibly be necessary, he now vehemently affirmed that if I had but taken him as guide, he would have let me down into holes 40 mètres deep, where I should have seen such things as man had never seen before. Had monsieur seen the source of the Loue? Yes, monsieur had. Very fine, was it not? Yes, very fine. Which did monsieur then prefer--the glacière, or the source? The source, infinitely. *Then* it was clear monsieur had not seen the glacière:--he was sure before that monsieur had not, *now* it was quite clear, for in all the world there was nothing like that glacière. The Loue!--one might rather see the glacière once, than live by the source of the Loue all the days of one's life.

It was now five o'clock, and the train left Pontarlier at half-past seven. We represented to M. Paget that he really ought to do the twenty kilomètres in two hours and a quarter, which would leave us a quarter of an hour to arrange our knapsacks and pay the *National*. He promised to do his best, and certainly the black horse proved himself a most willing beast.

There was one long hill which damped our spirits, and made us give up the idea of catching the train; and here our driver came to the rescue with what sounded at first like a promising story--the only one we extracted from him all through the day--*à propos* of a memorial-stone on the road-side, where a man had lately been killed by two bears; but, when we came to examine into it, the romance vanished, for the man was a brewer's waggoner with a dray of beer, and the bears were tame bears, led in a string, which frightened the brewer's horses, and so the man was killed. Contrary to our expectations and fears, we did catch the train, and arrived in a thankful frame of mind at comfortable quarters in Neufchâtel.

Chapter 9

The Schafloch, or Trou-Aux-Moutons, Near the Lake of Thun

The next morning, my sisters went one way and I another; they to a valley in the south-west of Vaud, where our head-quarters were to be established for some weeks, and I to Soleure, where a Swiss *savant* had vaguely told us he believed there was a glacière to be seen. That town, however, denied the existence of any approach to such a thing, with a unanimity which in itself was suspicious, and with a want of imagination which I had not expected to find. One man I really thought might be persuaded to know of some cave where there was or might be ice, but after a quarter of an hour's discussion he finally became immovable on the negative side. A Frenchman would certainly have been polite enough to accommodate facts to my desires. It was all the more annoying, because the Weissenstein stood overhead so engagingly, and I should have been only too glad to spend the night in the hotel there, if anyone had given me the slightest encouragement. I specially pointed at the neighbourhood of this hotel to my doubtful friend, as being likely for caves; but he was not in the pay of the landlord, and so failed to take the hint. There is a curious hole in which ice is found near Weissenstein in Carniola,[55] and it is not

[55] See p. 258.

impossible that this may have originated the idea of a glacière near Soleure.

The Schweizerhof at Berne is a very comfortable resting-place; but, in spite of its various excellences, if a tired traveller is told that No. 53 is to be his room, he will do well to seek a bed elsewhere. No. 53 is a sort of closet to some other number, with a single window opening low on to the passage, and is adjudged to the unfortunate individual who arrives at that omnipresent crisis which raises the charge for bed-rooms, and silences all objections to their want of comfort--namely, when there is only one bed left in the house. In itself, No. 53 would be well enough; but the throne of the chambermaid is in the passage, by the side of the window, and the male attendant on that particular stage naturally gravitates to the same point, when the bells of the stage do not summon him elsewhere, and often enough when they do. This combination leads of course to local disturbances of a somewhat noisy character, and however entirely a sleepy man may in principle sympathise with the causes of the noise, it becomes rather hard to bear after midnight. The precise actors on the present occasion have, no doubt, quarrelled or set up a *café* before now, or perhaps have achieved both results by taking the latter first; but there is reason to believe that so long as the window of No. 53 is the seat of the chambermaid for the time being, so long will that room be--as the landlord neatly expressed it when a protest was made--*etwas unruhig*.

All Switzerland has been playing at soldiers for some time, and as we left Berne the next morning, we saw three or four hundred Federal men of war marching down the road which runs parallel with the rails. The three officers at the head of the column were elderly and stout; moreover, they were mounted, and that fact was evidently due rather to the meekness of their chargers than to the grip of their own legs. When they saw the train coming, they took prompt measures. They halted the troops, and rode off down a side lane to be out of harm's way; and when we had well passed, they rejoined the column, and the march was resumed.

The early train from Berne catches the first boat on the Lake of Thun, and I landed at the second station on the lake, the village of Gonten or Gunten. M. Thury's list states that the glacière known as the Schafloch is on the Rothhorn, in the Canton of Berne, 4,500 mètres of horizontal distance from Merligen, a village on the shore of the lake; and from these data I was to find the cave. Gonten was apparently the nearest station to Merligen, and as soon as the small boat which meets the steamer had deposited me on the shore, I asked my way, first to the *auberge*, and then to Merligen. The *auberge* was soon found, and coffee and bread were at once ordered for breakfast; but when the people learned my eventual destination, they would not let me go to Merligen. A man, to whom--for no particular reason--I had given two-pence, called a council of the village upon me, and they proceeded to determine whether I must have a guide from Gonten, or only from a nameless châlet higher up. The discussion was noisy, and was conducted without words: they do not speak, those men of Gonten--they merely grunt, and each interprets the grunts as he wills. My two-penny friend told me what it all meant, in an obliging manner, but in words less intelligible than the grunts; and one member of the council drew out so elaborate a route--the very characters being wild patois--splitting the morning into quarter-stundes and half-quarter-stundes, with a sharp turn to the right or left at the end of each, that, as I drank my coffee, I determined to take a guide from the village, whatever the decision of the council might be. Fortunately, things took a right turn, and when breakfast was finished, a deputation went out and found a guide, suspiciously like one of their number who did not return, and I was informed that Christian Opliger would conduct me to the Schafloch for five francs, and a *Trinkgeld* if I were satisfied with him. In order to prove to me that he had really been at the cave, six days before, with two Bernese gentlemen, he seized my favourite low-crowned white hat, and endeavoured to knead it into the shape of the cave.

Our affairs took a long time to arrange, for grunts and pantomime are not rapid means of communication, when it comes to detail. The great question in Christian's mind seemed to be, what should we take with us to eat and drink? and when he propounded this to me with steady pertinacity, I, with equal pertinacity, had only one answer--a cord and a hatchet. At last he provided these, vowing that they were ridiculously unnecessary, but comprehending that they must be forthcoming, as a preliminary to anything more digestible; and then I told him, some dry bread and no wine. This drove him from grunts to words. No wine! it would be so frightfully hot on the mountains!--I told him I never drank wine when I was hot. But it would be so terribly cold in the cave!--I never drank wine when I was cold. But the climbing was *sehr stark*--we should need to give ourselves strength!--I never needed to give myself strength. There was no good water to be found the whole way!--I never drank water. Then, at last, after a brief grunt with the landlord, he struck:--he simply would not go without wine! I never wished him to do so, I explained; he might take as much as he chose, and I would pay for it, but he need not count me for anything in calculating how much was necessary. This made him perfectly happy; and when I answered his question touching cheese in a similar manner, only limiting him to a pound and a half, he rushed off for a large wicker *hotte*, spacious enough for the stowage of many layers of babies; and in it he packed all our properties, and all his provisions. The landlord had made his own calculations, and put it at 3lbs. of bread and 2lbs. of cheese; but I cut down the bread on account of its bulk, before I saw the size of the *hotte*, and Christian seemed to think he had quite enough to carry.

It was about half-past nine when we started from the *auberge*; and after a short mount in the full sun, we were not sorry to reach the pleasant shade of walnut trees which accompanied us for a considerable distance. The blue lake lay at our feet on the right, and beyond it the Niesen stood, with wonted grandeur, guarding its subject valleys; more in front, as we ascended transversely, the well-known snow-peaks of the Bernese Oberland glittered high above the nearer foreground, and, sheer above us, on the left, rose the ragged precipices whose flank we were to turn. The Rothhorn of the Canton Berne lies inland from the Lake of Thun, and sends down towards the lake a ridge sufficiently lofty, terminating in the Ralligstöcke, or Ralligflue, the needle-like point, so prettily ridged with firs, which advances its precipitous sides to the water. These precipices were formed in historic times, and the sheer face from which half a mountain has been torn stands now as clear and fresh as ever, while a chaos of vast blocks at its foot gives a point to the local legends of devastation and ruin caused by the various berg-falls. Two such falls are clearly marked by the *débris*: one of these, a hundred and fifty years ago, reduced the town of Ralligen to a solitary Schloss; and the other, in 1856, overwhelmed the village of Merligen, and converted its rich pastures into a desert cropped with stones. A traveller in Switzerland, at the beginning of this century, found that the inhabitants of Merligen were considered in the neighbourhood to be *d'une stupidité et d'une bêtise extrêmes*, and I am inclined to believe that after the last avalanche a general migration to Gonten must have taken place.

Christian's patois was of so hopeless a description, that I was tempted to give it up in despair, and walk on in silence. Still, as we were together for a whole long day, for better or for worse, it seemed worthwhile to make every effort to understand each other, else I could learn no local tales and legends, and Christian would earn but little *Trinkgeld*; so we struggled manfully against our difficulties. A confident American lady, meditating Europe, and knowing little French and no German, is said to have remarked jauntily that if the worst came to the worst she could always talk on her fingers to the peasants; but I did not attempt to avail myself of the results of early practice in that universal language. Christian's answers--the more intelligible parts of them--were a stratified succession of *yes* and *no*, and as he was a man naturally polite and acquiescent, the assentient strata were of more frequent occurrence; but of course, beyond showing his good-will, such answers were of no practical value. At length, after long perseverance, we were rewarded by the appearance of a curiosity which eventually gave each the key to the other's cipher. This was a strong stream of water, flowing out of the trunk of a growing tree, at a height of six feet or so from the ground; and I was so evidently interested in the phenomenon, that Christian exerted himself to the utmost, at last with success, to explain the construction of the fountain. A healthy poplar, seven or eight years old, is taken from its native soil, and a cold iron borer is run up the heart of the trunk from the roots, for six feet or more, by which means the pith is removed, and the trunk is made to assume the character of a pipe. A hole is then bored through from the outside of the trunk, to communicate with the highest point reached by the former operation, and in this second hole a spout is fixed. The same is done at a very short distance above the root, in the part of the trunk which will be buried in the earth when the tree is replanted, and the poplar is then fixed in damp ground, with the pipe at its root in connection with one of the little runs of water which abound in meadows at the foot of hills. A well-known property of fluids produces then the strange effect of an unceasing flow of water from an iron spout in the trunk of a living tree; and, as poplars love water, the fountain-tree thrives, and is more vigorous than its neighbours. This sort of fountain may be common in some parts of Switzerland, but I have not seen them myself except in this immediate neighbourhood. There is said to be one near Stachelberg.

In the endeavour to explain all this to me, Christian succeeded so perfectly, that for the rest of the day we understood each other very well. When I told him that he spoke much better German than the rest of the people in Gonten, he informed me that he had worked among foreigners, in proof whereof he held out his fingers; but all that I could gather from the invited inspection was, that, whatever his employment might have been, he could not be said to have come out of it with clean hands. He had been employed, he explained, in German dye-works, and there had learned something better than the native patois. About this time, too, I was able to make him understand that, as he carried more than I, he must call a halt whenever he felt so inclined; upon which he patted me affectionately on the back, and, if I could remember the word he used, I believe that I should now know the Swiss-German for a brick.

Our object was to pass along the side of the lake, at a considerable elevation, till we reached the east side of the Rothhorn range, when we were to turn up the Jüstisthal, and mount towards the highest point of the ridge, the glacière lying about an hour below the summit, in the face of the steep rock. The cliffs became very grand on either side, as soon as we entered this valley, the Jüstisthal, especially the precipices of the Beatenberg on the right; and our path lay through woods which have sprung up on the site of an early *Berg-lauine.* The guide-books call attention to a cavern with a curious intermittent spring in this neighbourhood. English tourists should feel some interest in the Cave of S. Beatus, inasmuch as its canonised occupant went from our shores to preach the Gospel to the wild men of the district, and died in this cave at a very advanced age. His relics remaining there, his fête-day attracted such crowds of pilgrims, that reforming Berne sent two deputies in 1528 to carry off the saint's skull, and bury it between the lakes; but still the pilgrimages continued, and at length the Protestant zeal of Berne went to the expense of a wall, and they built the pilgrims out in 1566. S. Beatus is said to have been converted by S. Barnabas in Britain, and to have gone to Rome, whence S. Peter sent him out to preach. His relics were conveyed to Lucerne in 1554, because heresy prevailed in the country where his cave lies, and an arm is among the proud possessions of pilgrim-pressed Einsiedeln. The saint was originally a British noble, by name Suetonius; and Dempster drops a letter from his name, and with much ingenuity makes him collateral ancestor of a Scottish family--'The Setons, tall and proud.'[56]

When we arrived at the last châlet, Christian turned to mount the grass slope on our left hand, which led to the part of the rocks in which the entrance to the Schafloch was to be sought. I never climbed up grass so steep, and before we had gone very far we were hailed by a succession of grunts, which my companion interpreted into assurances from some invisible person that we were going wrong. The man soon appeared, in the shape of a charcoal-burner, and told us that we were making the ascent much more difficult than it need be made, and also, that we should come to some awkward rock-climbing by the route we had chosen. It was too late, however, to turn back; so we persevered.

Before long, I heard a *Meinherr*! from Christian, in a tone which I knew meant rest and some food. He explained that he would rather take two small refreshments, one here and one at the Schafloch, than one large

---

[56] Acta SS. Bolland. May 9.—If possessed of the characteristics of his race—'tall and proud'—his activity belies the first line of the old saying,

'Lang and lazy,
Little and loud;
Red and foolish,
Black and proud:'

though possibly the personal habits which a modern spirit loves to point out, as the great essential of hermit-life, united with the family characteristic of the early Seton to verify the last line of the saying.

refreshment at the cave; so we propped ourselves on the grass, and tapped the *hotte*. The cheese proved to be delightful--six years old, the landlady told us afterwards, and apparently as hard as a bone, but when once mastered its flavour was admirable. Christian persuaded me to taste the wine, of which he had a high opinion, and he was electrified by the universal shudder the one taste caused. The grapes from which it was brewed had been grown in a gooseberry garden, and all the saccharine matter carefully extracted; the wine had been left without a cork since the first dawn of its existence, and the heat and jolting of its travels on Christian's back had reduced it to the condition of warm flat vinegar. He drank it with the utmost relish, and was evidently reconciled to my verdict by the consideration that there would be all the more for him.

From the appearance of the bread and cheese when the meal had come to an end, I concluded that my companion had changed his mind in the course of feeding, and had resolved to compress the whole eating of the day into one large refreshment here. The consumptive powers of the Swiss-German peasant, when his meal is franked, has not unfrequently reminded me of the miraculous eating performed by a yellow domino of that nation, at the fête by which Louis XIV celebrated the second marriage of the Dauphin. This domino was of large size, and ate and drank voraciously throughout the entertainment, which lasted many hours, retiring every five minutes or so, and returning speedily with unabated appetite. The thing became at length so portentous, that enquiries were instituted, and it was found that the trusty *Cent-Suisses* had joined at a domino, and were drawing lots all through the evening for the next turn at eating; so that each man's time was necessarily limited, and he accordingly made the most of it.

We soon took to the rocks, and found them, as the charcoal-burner had promised, sufficiently stiff work. Colonel (now General) Dufour visited the Schafloch with a party of officers in 1822, and he describes[57] the path as a dangerous one, so much so that several of the gallant members of his party could not reach the cave: he uses rather large words about the precipices, and it is a matter of observation that military service on the Continent tends

[57] Bibl. Univ. de Genève, First Series, xxi. 113. See also Edinburgh Philosophical Journal, viii. 290.

to induce a habit of body which is not the most suitable for doubtful climbing. The mountain seemed to be composed, in this part, of horizontal layers of crumbling shale, with a layer now and then of stone, about the thickness of an ordinary house-tile. The stone layers project from the looser masonry, and afford an excellent foot-hold; but a slip might be unpleasant. Everyone who has done even a small amount of climbing has met with an abundance of places where 'a slip would be certain death,' as people are so fond of saying; but equally he has discovered that a slip is the last thing he thinks of making in such situations. Christian had told me that if I had the slightest tendency to *Schwindelkopf*, I must not go by the improvised route; but it proved that there were really no precipices at all, much less any of sufficient magnitude to turn an ordinary head dizzy. He chose these rocks as the text for a long sermon on the necessity for great caution when we should arrive at the cave, telling of an Englishman who had tried to visit it two years before, and had cut his knee so badly with his guide's axe that he had to be carried down the mountain to Gonten, and thence to the steamer for Thun, in which town he lay for many weeks in the hands of the German doctor; this last assertion being by no means incredible. Also, of a native who attempted the cave alone, and, making one false step near the top of a fall of ice, slipped down and down almost forever, and finally landed with broken limbs on a floor of ice, where he was found, two days after, frozen stiff, but still alive.

It was not necessary to mount much, for we were almost as high as the mouth of the cave, according to Christian's belief, and our work consisted chiefly in passing along the face of the rock, round projecting buttresses and re-entering angles, till we reached that part of the mountain where we might expect to find our glacière. While we were thus engaged, two hoarse and ominous ravens took us under their charge, and accompanied us with unpleasant screams, which argued the proximity of food or nest. We soon found that we had disturbed their meal, for we came to marks of blood, and saw that some animal had slipped on the rocks above, and landed on the ledge on which we were walking, bounding off again on to a shelf below, where the ravens had already torn the body to pieces. I must confess to a very considerable shudder when we discovered the reason of their screams, and neither of us seemed to enjoy the circling and croaking of the unclean birds.

Very soon after this, Christian announced that we had reached the cave, and a steep little climb of six feet or so brought us to the entrance. Here we were haunted still by the presence of pieces of the fallen goat, which lay about here and there on the ground; and the flutter of wings overhead explained to us that the old ravens had built their nest in the mouth of the cave, and had brought morsels of raw flesh to their young ones, which were scarcely able to fly. I am ashamed to say that we were so angry with the old birds for shrieking so suggestively in our ears, and parading before us the results of a slip on the rocks, that we charged ourselves with stones, and put an end to the most noisy member of the foul brood; Christian making some of the worst shots it is possible to conceive, and raining blocks of stone and lumps of wood in all directions, with such reckless impartiality, that the only safe place seemed to be between him and the bird. One of us, at least, regretted the useless cruelty as soon as it was perpetrated, and it came back upon me very reproachfully at an awkward part of our return journey.

The Schafloch does not take its name from the bones contained in it, as is the case with the Kühloch in Franconia,[58] but from the fact that when a sudden storm comes on, the sheep and goats make their way to the cave for shelter, never, I was told, going so far as the commencement of the ice. The entrance faces ESE, and is of large size, with a low wall built partly across it to increase the shelter for the sheep: Dufour calls the entrance 50 feet wide and 25 feet high, but I found the width at the narrowest part, a few yards within the entrance, to be 33 feet.[59] For a short distance the cave passes horizontally into the rock, in a westerly direction, and is quite light; it then turns sharp to the south, the floor beginning to fall, and candles becoming necessary. Here the height increases considerably, and the way lies over a wild confusion of loose masses of rock, which have apparently

[58] Philosophical Magazine, Aug. 1829.

[59] Colonel Dufour guessed the elevation of the cave, in 1822, at two-thirds the height of the Niesen, and forty years after, as General Dufour, he published the result of the scientific survey of Switzerland, which makes it 1,780 mètres; so that his early guess was not a bad one.

fallen from the roof, and make progression very difficult. We soon reached a point where ice began to appear among the stones; and as we advanced it became more and more prominent, till at length we lost sight of the rock, and stood on solid ice.

On either side of the cave was a grand column of ice forming the portal, as it were, through which we must pass to further beauties. The ice-floor rose to meet these columns in a graceful swelling curve, perfectly continuous, so that the general effect was that of two columns whose roots expanded and met in the middle of the cave; and, indeed, that may have been really the order of formation. The right-hand column was larger than its fellow, but, owing to the more gradual expansion of the lower part of its height, and the steepness of the consequent slope, we were unable to measure its girth at any point where it could be fairly called a column. Christian had been in the cave a few days before, and he assured me that the swelling base of this column had increased very considerably since his last visit, pointing out a solid surface of ice, at one part of our track, where he had before walked on bare rock. The cave was by no means extremely cold, that is to say, it was rather above than below the freezing point, and the splashing of drops of water was audible on all sides; so that, if Christian spoke the truth,--it was sad to be so often reminded of Legree's plaintive soliloquy in the opening pages of 'Uncle Tom's Cabin,'--the explanation, I suppose, might be that the drops of water, falling on the top of the column or stalagmite, run down the sides, and carry with them some melted portion from the upper part of the column, and after a course of a few yards become so far refrigerated as to form ice.[60] The pillar on the left was more approachable, but we were unable to determine its dimensions; for on the outer side, where it stood a few feet or yards clear of the side of the cave, the rounded ice at its foot fell off at once into a dark chasm, a sort of smooth enticing *Bergschrund*, which we did not care to face. Christian declared that this column was not so high as it was a day or two before, which may go to support the theory expressed above, or at least that part of

---

[60] There is a hint of something of this kind in an editorial note in the *Journal des Mines* (now *Annales des Mines*) of Prairial, an. iv. pp. 71, 72, in connection with the glacière near Besançon.

it which depends upon the supposition of water dropping on to the head of the column, and melting certain portions of it.

If we were unable to take the external dimensions of this column, I had no doubt that we should find internal investigations interesting; so, to Christian's surprise, I began to chop a hole in it, about two feet from the ground, and, having made an entrance sufficiently large, proceeded to get into the cavity which presented itself. The flooring of the dome-shaped grotto in which I found myself, was loose rock, at a level about two feet below the surface of the ice-floor on which Christian still stood. The dome itself was not high enough to allow me to stand upright, and from the roof, principally from the central part, a complex mass of delicate icicles passed down to the floor, leaving a narrow burrowing passage round, which was itself invaded by icicles from the lower part of the sloping roof, and by stubborn stalagmites of ice rising from the floor.[61] The details of this central cluster of icicles, and in fact of every portion of the interior of the strange grotto, were exceedingly lovely, and I crushed with much regret, on hands and knees, through fair crystal forests and frozen dreams of beauty. In making the tour of this grotto, contorting my body like a snake to get in and out among the ice-pillars, and do as little damage as might be, but yet, with all my care, accompanied by the incessant shiver and clatter of breaking and falling ice, I came to a hole in the ground, too dark and deep for one candle to show its depth; so I called to Christian to come in, thinking that two candles might show it better. He asked if I really meant it, and assured me he could be of no use; but I told him that he must come, and informed him that he, being the smaller man, would find the passage quite easy. It was very fortunate that I had not waited a minute longer before summoning him, for just as he had dropped into the hollow, and was beginning his journey to the side where I now was, a drop of water and a simultaneous icicle came upon my candle, and left me in darkness, curled up like a dormouse in a nest of ice, at the edge of the newly discovered shaft; while my troubles were brought to a climax by an incursion of icy

[61] M. Soret, who visited the Schafloch in September 1860, and communicated his notes to M. Thury, speaks of many columns in this part of the glacière, where we found only two. 'L'un d'entre eux,' he says, 'présentait dans sa partie inférieure une petite grotte ou cavité, assez grande pour qu'un homme pût y entrer en se courbant.'

drops, which had me at their mercy. If all this had happened while Christian was still outside, he would probably have staid there wringing his hands till it was time to go home, and I should certainly not have liked to move without a light. As it was, I did not inform him of the catastrophe, but let him come toiling on, wondering audibly what madness could drive Herrschaft into such places; and when he arrived, we cut off the wet wick, and lighted the candle again. We could make nothing of the hole, so he returned by the way he had come, and I completed the tour of the grotto, finding the same difficult passage, and the same ice beauties, all the way round.

Having squeezed ourselves out again through the narrow hole, we now passed between the two gigantic columns, and found that the sea of ice became still broader and bolder. I much regret that I neglected to take any measurements in this part of the cave; but farther down, where it was certainly not so broad, I found the width of the ice to be 75 feet. It was throughout of the crystalline character which prevails in all the large masses in the glacières I have visited. For some distance beyond the columns, we found neither stalactites nor stalagmites--indeed, I forgot to look at the roof--until we came to the edge of a glorious ice-fall, down which Christian said it was impossible to go--no one had ever been farther than where we now stood. I have seen no subterranean ice-fall so grand as this, round and smooth, and perfectly unbroken, passing down, like the rapids of some river too deep for its surface to be disturbed, into darkness against which two candles prevailed nothing. The fall in the Upper Glacière of the Pré de S. Livres was strange enough, but it was very small, and led to a confined corner of the cavern; whereas this of the Schafloch rolls down majestically, cold and grey, into a dark gulf of which we could see neither the roof nor the end, while the pieces of ice which we despatched down the steep slope could be heard going on and on, as M. Soret says, *à une très-grande distance*. The shape, also, of the fall was very striking. Beginning at the left wall of the cave, the edge ran out obliquely towards the middle, when it suddenly turned and struck straight across to the right-hand wall, so that we were able to stand on a tongue, as it were, in the middle of the top of the fall. To add to the effect, precisely from this tongue or angle a fine column of ice sprang out of the very crest of the fall, rising to or towards the roof, and to this we clung to peer down into the darkness.

The rope we had brought was not long, and the idea was hopeless of cutting steps down this great fall, leading we knew not where, with an incline which it frightened Christian even to look at. I began to consider, however, whether it was not possible to make our way down the left branch of the ice, which fell rather towards the side wall than into the dark gulf below. On examining more closely, I found that a large stone, or piece of rock, projected from the face of this branch of the fall, about 12 feet from the top, and to this I determined to descend, as a preliminary to further attempts, the candles not showing us what there was beyond. Accordingly, I tied on the rope, and planted Christian where he had a safe footing, telling him to hold tight if I slipped, for he seemed to have little idea what the rope was meant for. The ice was very hard, and cutting steps downwards with a short axe is not easy work; so when I came within 3 or 4 feet of the rock, I forgot the rope, and set off for a short glissade. Christian, of course, thought something was wrong, and very properly put a prompt strain upon the rope, which reduced his Herr to a spread-eagle sort of condition, in which it was difficult to explain matters, so as to procure a release. When that was accomplished, I saw it would be easy to reach the point where the ice met the wall, so I called to Christian to come down, which he did in an unpremeditated, avalanche fashion; and then, by cutting steps here and there, and making use of odd points of rock, we skirted down the edge of the great fall, and reached at last the lower regions.

When I came to read Dufour's account of his visit in 1822, I found that the ice must have increased very much since his time. He uses sufficiently large words, speaking of the *vaste, horrible et pourtant magnifique*--of the *horreur du séjour*, and the *grandeur des demeures souterraines*; but he only calls the glorious ice-fall a *plan incliné*, and says that the whole was less remarkable for the amount of ice, than for the characteristics indicated by the words I have quoted. He says that it required *une assez forte dose de courage* to slip down to the stone of which I have spoken; the fact being that at the time of my visit it would have been impossible to do so with any chance of stopping oneself, for the flat surface of the stone was all but even with the ice. M. Soret, who saw the cave in 1860, determined that cords were then absolutely necessary for the descent, which he did not attempt; and the only Englishman I have met who has seen this cave, tells me that he and his party went no farther than the edge of the fall.[62] Probably each year's accumulation on the upper floor of ice has added to the height and rapidity of the fall; but at any rate, when Dufour was there, *des militaires*--as he dashingly tells--were not to be stopped, and he and his party--such of them as had not been already stopped by the precipices outside--let themselves slip down to the stone, and thence descended as we did.

We soon found that the larger ice-fall looked extremely grand when seen from below, and that in a modified form it reached far down into the lower cave, and terminated in a level sea of ice; but, before making any further investigations into its size, we pressed on to look for the end of the cave. This soon appeared, and as a commentary on Christian's assertion that no one had ever been beyond the head of the fall, I called his attention to some initials smoked on the wall by means of a torch. There was an abrupt piece of rock-floor between this end and the termination of the ice. The western wall was ornamented with a long arcade of lofty columns of very white ice, looking strangely ghostlike by the light of two candles, crystallised, and with the porcelain appearance I have described before. We could not measure the height of these columns, but we found that they extended continuously, so as to be in fact one sheet of columns, connected by shapes of ice now graceful and now grotesque, for 27 yards. The ice from their feet flowed down to join the terminal lake, which formed a weird sea 28 yards by 14. My notes, written on the spot, tell me that between this lake, which I have called terminal, and the end of the cave, there is a sheet of ice 48 yards long, but it has entirely vanished from my recollection.

I now sent Christian back with a ball of string, up the steps we had cut for the descent, with directions to get as near as he could to the top of the main fall, and then send down a stone tied to the string, as I wished to

---

[62] See also the note at the end of this chapter.

determine the length of the fall. While he was making his way up, I amused myself by chopping and carving at the ice at various points to examine its structure, until at length a *Jodel* from above announced that Christian had reached his post; and a vast amount of hammering ensued, of which I could not understand the meaning. Presently he called out that 'it' was coming, and assuredly it did come. There was a loud crash on the upper part of the fall, and a shower of fragments of ice came whizzing past, and almost dislodged me; while the sound of pieces of ice bounding and gliding down the slope seemed as if it never would cease. It turned out to mean that my friend had not been able to find a stone; so he had smashed a block of ice from the column which presided over the fall, and having attached the string to this, had hurled the whole apparatus in my direction, fortunately not doing as much damage as he might have done. My end of the string was not to be seen, so he repeated the experiment, with a piece of wood in place of the block of ice, and this time it succeeded. We found that from top to bottom of the fall was 45 yards. There was all the appearance of immense thickness, especially towards the upper part.

Christian had placed his candle in a niche in the column, while he arranged the string for measuring the fall, and the effect of the spark of light at the top of the long steep slope was extremely strange from below. The whole scene was so remarkable, that it required some effort to realise the fact that I was not in a dream. Christian stood at the top invisible, jodeling in a most unearthly manner, and developing an astonishing falsetto power, only interrupting his performance to assure me that he was not coming down again; so I was obliged to measure the breadth of the fall by myself. I chose a part where the ice was not very steep, and where occasional points of rock would save some of the labour of cutting steps; but even so it was a sufficiently tedious business. The string was always catching at something, and mere progression, without any string to manage, would have been difficult enough under the circumstances. It was completely dark, so a candle occupied one hand, and, as every step must be cut, save where an opportune rock or stone appeared, an axe occupied the other; then there was the string to be attended to, and both hands must be ready to clutch at some projecting point when a slip came, and now and then a ruder rock required circumvention. Add to all this, that hands and feet had not been rendered more serviceable by an hour and a half of contact with ice, and it will easily be understood that I was glad when the measurement was over. At this point the breadth was 25 yards, and, a few feet above the line in which I crossed, all traces of rock or stone disappeared, and there was nothing but unbroken ice. I had of course abundant opportunities for examining the structure of the ice, and I found in all parts of the fall the same large-grained material, breaking up, when cut, into the usual prismatic nuts.

I now rejoined Christian, and we worked our way upwards to the mouth of the cave, penitently desisting from stoning a remaining raven. We observed at the very mouth, by watching the flame of the candles, a slight current outwards, extremely feeble, and on our first arrival I had fancied there was a current, equally slight, inwards, but neither was perceptible beyond the entrance of the cave. M. Soret was fortunate enough to witness a curious phenomenon, at the time of his visit to the Schafloch, in September 1860, which throws some light upon the atmospheric state of the cave. The day was externally very foggy, and the fog had penetrated into the cavern; but as soon as M. Soret began to descend to the glacière itself, properly so called, he passed down out of the fog, and found the air for the rest of the way perfectly clear.[63]

M. Soret states that he has not absolute confidence in his thermometrical observations, but as he had more time than I to devote to such details, inasmuch as he did not pass down into the lowest part of the cave, I give his results rather than my own, which were carelessly made on this occasion:--On a stone near the first column of ice, 0°·37 C.; on a stick propped against the column on the edge of the great ice-fall, 2°·37 C.; in a hole in the ice, filled with water by drops from the roof, 0° C. approximately.[64]

---

[63] 'Toute la couche supérieure au plan de niveau passant par le seuil était chargée de brouillard; toute la couche inférieure à ce niveau était parfaitement limpide.' (*Thury*, p. 37.)

[64] Respectively, 32°·666, 36°·266, and 32°, Fahrenheit.

The second result is sufficiently remarkable. My own observations would give nearer 33° F. than 32° as the general temperature of the cave.

Christian was so cold when we had finished our investigations, that he determined to take his second refreshment *en route*, and, moreover, time was getting rather short. We had started from Gonten at half-past nine in the morning, and reached the glacière about half-past twelve. It was now three o'clock, and the boat from Gonten must reach the steamer at half-past six precisely, so there was not too much time for us; especially as we were to return by a more mountainous route, which involved further climbing towards the summit of the Rothhorn, and was to include a visit to the top of the Ralligflue. On emerging from the cave, we were much struck by the beauty of the view, the upper half of the Jungfrau, with its glittering attendants and rivals, soaring above a rich and varied foreground not unworthy of so glorious a termination. There was not time, however, to admire it as it deserved, and we set off almost at once up the rocks, soon reaching a more elevated table-land by dint of steep climbing. The ground of this table-land was solid rock, smoothed and rounded by long weathering, and fissured in every direction by broad and narrow crevasses 2 or 3 feet deep, at the bottom of which was luxuriant botany, in the shape of ferns, and mallows, and monkshood, and all manner of herbs. The learned in such matters call these rock-fallows *Karrenfelden*. When we had crossed this plateau, and came to grass, we found a gorgeous carpet of the huge couched blue gentian (*G. acaulis*, Fr. *Gentiane sans tige*), with smaller patterns put in by the dazzling blue of the delicate little flower of the same species (*G. verna*); while the white blossoms of the grass of Parnassus, and the frailer white of the *dryade à huit petales*, and the modest waxen flowers of the *Azalea procumbens* and the *airelle ponctuée* (*Vaccineum vitis idaea*), tempered and set off the prevailing blue. There were groves, too, rather lower down, of Alpine roses (the first I had come across that year), not the fringed or the green-backed species which botanists love best, but the honest old rust-backed rhododendron, which every Swiss traveller has been pestered with in places where the children are one short step above mere mendicity, but, equally, which every Swiss traveller hails with Medean delight when he comes upon it on the mountain-side. We were now, too, in the neighbourhood of the first created Alpen rose. The story is, that a young peasant, who had climbed the precipices behind Oberhausen for rock-flowrets, as the price of some maiden's love, fell at the moment when he had secured the flowers, and was killed. From his blood the true Alpen rose sprang, and took its colour.

We were now passing along the summit of one of the lower spurs of the Rothhorn range, and making for the peak of the Ralligflue, which lay considerably below us. In descending near the line of crest, we found a large number of very deep fissures, narrow and black, some of them extending to a great distance across the face of the hill; sometimes they appeared as mere holes, down which we despatched stones, sometimes as unpleasant crevasses almost hidden by flowers and the shrubs of rhododendron. In many of these we dimly discovered accumulated snow at the bottom, and we observed that the Alpine roses which overhung the snow-holes were by far the deepest coloured and most beautiful we could find.

To reach the Ralligflue, we had to cross a smooth green lawn completely covered with the sweet vanilla orchis (*O. nigra*), which perfumed the air almost too powerfully. No one can ever fully appreciate the grandeur of the lion-like Niesen till he has seen it from this verdant little paradise, on the slope near the Bergli Châlet, with a diminutive limpid lake in the meadow at his feet, and the blue lake of Thun below. The Kanderthal and the Simmenthal lie exposed from their entrance at the foot of the Niesen; and when the winding Kanderthal is lost, the Adelbodenthal takes up the telescope, and guides the eye to the parent glaciers. This view I was fortunately able to enjoy rather longer than that from the mouth of the Schafloch; for we had made such rapid way, that Christian found there was time for a meal of milk in the châlet, and meanwhile left me lying in perfect luxury on the sweet grass.

From the Ralligflue a long and remarkably steep zigzag leads to the lower ground, and down this Christian ran at full speed, jodeling in a most trying manner; indeed, at one of the sudden turns of the path he went off triumphantly into a falsetto so unearthly, that he lost his legs, and landed in a promiscuous sort of way on a lower part of the zigzag, after which he was slower and less vocal.

We eventually reached Gonten so soon, that there was time to cool and have a bath in the lake; and when that was nearly finished, Christian brought a plate of cherries and a detachment of the village, and I ate the cherries and held a levée in the boat--very literally a levée, as the dressing was by no means accomplished when the deputation arrived. My late guide, now, as he said, a friend for life, made a speech to the people, setting forth that he had done that day what he had never thought to do; for, often as he had been to the entrance of the Schafloch--five or six times at the least--he had never before reached the end of the cave. And to whom, he asked, did he owe it? All previous Herrschaft under his charge had cried *Immer zurück!* but this present Herr had known but one cry, *Immer vorwärts!* Luckily the steamer now approached, so the speech came to an end, and he shook hands affectionately, with a vigour that would certainly have transmitted some of the dye, if that material had not become a part of the skin which it coloured. Then the village also shook hands, having evidently understood what Christian said, notwithstanding the fact that it was intelligible German, and I returned to Thun and Berne.

No. 53 was still the only bed disengaged, for it was very late when I reached Berne; but on my vehement protestations against that unquiet chamber, the landlord most obligingly converted a sofa in his own sitting-room into a temporary bed, and made it over to me. This room was separated by a door of ground-glass from another sitting-room brilliantly lighted, in which a number of German young gentlemen were fêting the return of a comrade after the national manner. The landlord said he thought it must soon be over, for he doubted whether they could last much longer; but their powers of endurance were greater than he had supposed. It will readily be imagined that German songs with a good chorus, the solo parts being very short, and received with the utmost impatience by the chorus, were even less soporific in their effect than the flirtations--though boisterous beyond all conventional propriety--of German housemaids and waiters.[65]

[65] Since I wrote this chapter, my attention has been called to a tourist's account of the Schafloch in *Once a Week* (Nov. 26, 1864), in an article called *An Ice-cavern in the Justis-Thal.* The writer says—'We proceeded to the farther end of the cavern, or at least as far as we thought it prudent, to ascertain where the flooring of ice rounded off into the abyss of unfathomable water we heard trickling below.' One of the party 'having taken some large stones with him, he began hurling them into the profound mystery. Presently a heavy double-bass gurgle issued forth with ominous depth of voice, indicating the danger of farther progress. Having thus ascertained that if either of us ventured farther he would most probably not return by the way he went, the signal of retreat was given, and in about forty minutes, after encountering the same amusing difficulties which had enlivened our descent, Æneas-like we gained the upper air.' It will be seen from my account of what we found in the 'abyss of unfathomable water,' that a little farther exploration might have effected a change in the writer's views.

Chapter 10

THE GLACIÈRE OF GRAND ANU, ON THE MONTAGNE DE L'EAU, NEAR ANNECY

M. Thury's list contained a bare mention of two glacières on the M. Parmelan, near Annecy, without any further information respecting them, beyond the fact that they supplied ice for Lyons. Their existence had been apparently reported to him by M. Alphonse Favre, but he had obtained no account of a visit to the caves. Under these circumstances, the only plan was to go to Annecy, and trust to chance for finding someone there who could assist me in my search.

After spending a day or two in the library at Geneva, looking up M. Thury's references, with respect to various ice-caves, and trying to discover something more than he had found in the books there, I started for Annecy at seven in the morning in the banquette of the diligence. On a fresher day, no doubt the great richness of the orchards and corn-fields would have been very striking; but on this particular morning the fields were already trembling with heat, and the trees and the fruit covered with dust; and there was nothing in the grouping of the country through which the road lay to refresh the baked and half-choked traveller. The voyage was to last four and a half hours, and it soon became a serious question how far it would be possible to face the heat of noon, when the earlier morning was so utterly unbearable.

Before very long, a counter-irritant appeared in the shape of a fellow-traveller, whose luggage consisted of a stick and an old pair of boots. The man was not pleasant to be near in any way, and he was evidently not at all satisfied with the amount of room I allowed him. He kept discontentedly and doggedly pushing his spare pair of boots farther and farther into my two-thirds of the seat, and once or twice was on the point of a protest, in which case I was prepared to tell him that as he filled the whole banquette with his smell, he ought in reason to be satisfied with less room for himself; but instead of speaking, he brought out a tobacconist's parcel and began to open it. Tobacco-smoke is all very well under suitable circumstances, but it is possible to be too hot and dusty and bilious to be able to stand it, and I watched his proceedings with more of annoyance than of resignation. The parcel turned out, however, to be delightful snuff, tastefully perfumed and very refreshing; and the politeness with which the owner gave a pinch to the foreign monsieur, after apportioning a handful to the driver and conductor, won him a good three inches more of seat. The inevitable cigar soon came; but it was a very good one, and no one could complain: all the same, I could not help feeling a malicious satisfaction when the *douaniers* on the French frontier investigated the spare boots--guiltless, one might have thought, of anything except the extremity of age and dirt--and drew from them a bundle or two of smuggled cigars, the owner trying in vain to look as if he rather liked it.

The Hôtel de Genève is probably the least objectionable of the hotels of Annecy; but the Poste-bureau is at the Hôtel d'Angleterre, and it was much too hot for me to fight with the waiters there, and carry off my knapsack to another house. It is generally a mistake--a great mistake--to sleep at a house which is the starting-place and the goal of many diligences. All the night through, whips are cracking, bells jingling, and men are shouting hoarsely or blowing hoarser horns. Moreover, the Hôtel d'Angleterre had apparently needed a fresh coat of paint and universal papering for many years, and the latter need had at this crisis been so far grappled with that the old paper had been torn down from the walls and now lay on the various floors, while large pies of malodorous sizing had been planted at the angles of the stairs. The natural *salle-à-manger* was evidently an excellent room, with oleander balconies, but it was at present in the hands of joiners, and a card pointed the way to the 'provisionary *salle-à-manger'*--not a bad name for it--in the neighbourhood of the kitchen.

There was one redeeming feature. The people of the house were nice-looking and well-dressed. But experience has taught me to view such a phenomenon in French towns of humbler rank with somewhat mixed feelings. When the house is superintended with a keen and watchful eye by a young lady of fashionable appearance, who takes a personal interest in a solitary traveller, and suggests an evening's *course* on the lake, or a morning's drive to some good view, and makes herself most winning and agreeable; who takes the words, moreover, out of the mouth of a man meditating an ordinary dinner, and assures him that she knows exactly what he wants, and he shall be well satisfied, with a sisterly air that makes the idea of francs and sous not sordid only, but impossible; I have slowly learned to expect that this fashion and condescension will appear in the bill. Prettiness is a very expensive item in such a case; and as these three were all combined to a somewhat remarkable degree at the Hôtel d'Angleterre, the eventual bill made me angry, and I should certainly try the Hôtel de Genève on any future visit to Annecy.

The first thing to be done was to determine the position of the Mont Parmelan. I was prepared to find the people of the town denying the existence of such a mountain; but, as it was visible from the door of the hotel, they could not go quite so far as that. The small crowd at the door repudiated the glacières with one voice, and pointed out how unlikely it was that Lyons should be supplied with ice from Annecy; nevertheless, I continued to ask my way in spite of protestation, till at length a lame man passed by, who said monsieur was quite right--he himself knew two glacières on the Mont Parmelan very well. He had never seen either of them, but he knew them as well as if he had. It was useless to go to them now, he added, for the owners extracted all the ice early in the year, and stored it in holes in the lower part of the mountain. He had no idea by what route they were to be approached from Annecy, or on which side of the Mont Parmelan they lay.

I now looked on the local map, and determined that the best plan would be to take the Bonneville diligence as far as Charvonnaz, the point on the road which seemed to lie nearest to the roots of the Mont Parmelan, and then be guided by what I might learn among the peasants. Everyone said there was no chance of getting to anything by that means; but as the hotel people saw that it was of no use to deny the glacières any longer, they proposed to take me to a man who knew the M. Parmelan well, and could tell me all about it. This man proved to be a keeper of voitures,--an ominous profession under the circumstances,--and he assured me that I could make a most lovely *course* the next day, through scenery of unrivalled beauty; and he eloquently told on his fingers the villages and sights I should come to. I suggested--without in the least knowing that it was so--that the drive might be all very well in itself, but it would not bring me to the glacières; on which he assured me that he knew every inch of the mountain, and there was not such a thing as a glacière in the whole district. At this moment, a gentlemanlike man was brought up by the waiter, and introduced to me as a monsieur who knew a monsieur who knew the proprietor of one of the glacières, and would he happy to conduct me to this second monsieur: so, without any very ceremonious farewell to the owner of the proffered voiture, we marched off together down the street, and eventually turned into a *café*, whose master was the monsieur for whom we were in search. Know the glacière?--yes, indeed! he had ice from it one year every morning. His wife and he had made a *course* to the campagne of M. the Maire of Aviernoz, and he--the cafétier--had descended for miles, as it were, down and down, till he came to an underground world of ice, wonderful, totally wonderful: there he perceived so immense a cold, that he drank a bottle of rhoom--a whole bottle--and drank it from the neck, *à l'Anglaise*. And when they had gone so far that great dread came upon them, they rolled a stone down the ice, and it went into the darkness--boom, boom, boom,--and he put on a power of ventriloquism which admirably represented the strange suggestive sound. Hold a moment! had monsieur a crayon? Yes, monsieur had; so the things were impetuously swept off a round marble table, and the excited little man drew a fancy portrait of the glacière. The way to reach it? Go by diligence to Charvonnaz--exactly what I had determined upon--and walk up to Aviernoz, where his good friend the maire would make me see his beautiful glacière, through the means of a letter which he went to write. It was absurd to see this hot little man sign himself 'Dugravel, *glacier*,' that being the style of his profession, naturally recalling the contradictory conduct of the Latin noun *lucus*.

The bones of S. Francis of Sales lie in the church of S. François in Annecy, and I made a pilgrimage in search of them through very unpleasant streets. After a time, the Italian west front of the church appeared; but the main door led into a demonstrative bakery, and the door of the north aisle was obscured by oleanders and a striped awning, and over it appeared the legend, '*Entrée de l'Hôtel*.' As a man politely explained, they had built S. Francis another church, and utilised the old one. The town itself seemed to be of the squalid style of antiquity--old, no doubt, but very dirty. It is pervaded by streams, which crop up among the houses, and flow through dark alleys and vaulted passages, rarely coming into daylight, and suggesting all manner of dark crimes. The red-legged French kettledrums are, if possible, more insolent here than in other places, and it is evident that the dogs are not yet reconciled to the annexation, for the guard swept through the streets amid a perfect tornado of howls from the negligent scavengers of the place. For my own part, I was not pleased with the change of rule, when I found that since Annecy has become French, the *vin d'Asti* has become dear, as being now a foreign wine.

The diligence for Bonneville was to leave Annecy at half-past four in the morning; so I told them to call me at four, intending to breakfast somewhere on the way. But of course, when four o'clock came, I had to call myself, and in a quarter of an hour a knock at the door announced half-past four. I pounced upon the man, and remonstrated with him, but he assured me it did not matter; and when I reminded him that the diligence was to leave at half-past four, he observed philosophically that it was quite true, and I had better make haste, for the poste was very punctual. At the door of the bureau a loaded diligence stood, marked *Annecy--Aix*, and I asked had the Bonneville diligence gone? It did not go till six, the clerk told me; but I reminded him he had said half-past four when I asked him last night. Half-past four?--true, here was the carriage standing at the door. But that was for Aix, not Bonneville, I pointed out to him. Pardon--it was marked Aix, but was in fact meant for Bonneville.

The diligence reached the end of the by-road leading to Villaz in about half an hour, and all the fever of Geneva and Annecy seemed to fly away before the freshness of this green little lane, with clematis in full flower pervading the hedges, and huge clusters of young nuts peeping out, and promising later delights to fortunate passers-by. But, alas! the little lane soon came to an end, and as I faced the fields of corn up the mountain-side, the hot thunderous air came rolling down in palpable billows, and oppressive clouds took possession of the surrounding hills. Three-quarters of an hour brought me to Villaz, a close collection of houses on the hill-side, with arched stone gateways leading into the farmyards,--a fortified style of agricultural building which seems to prevail in that district. After an amount of experience in out-of-the-way places which makes me very cautious in saying that one in particular is dirtier than a dozen others, I venture to say that the *auberge* of Villaz is the most squalid I have come across; and I would not feed there again, except in very robust health, even for a new glacière. Still, it was absolutely necessary to eat something, and the landlady promised coffee and bread. She showed me first into the kitchen; but as it was also the place where the domestics slept, with many quadrupeds, I declined to sit there. Upon this she led me to the *salon*, where the window resisted all our efforts for some little time, and then opened upon such a choice assortment of abominations, that I fled without my baggage. The next attempt she made was the one remaining room of the house, the family bedroom; but that was so much worse than all, that I took final refuge on the balcony, a sort of ante-room to the hen-house. The cocks at the *auberge* of Villaz are the loudest, the hens the most talkative, and the cats the most shaggy and presuming, I have ever met with. Even here, however, all was not unmitigated darkness; for they ground the coffee while the water was boiling, and the consequent decoction was admirable. Moreover, the bread had a skin of such thickness and impervious toughness, that the inside was presumably clean.

Aviernoz lay about an hour farther. Almost as soon as I left Villaz, the thunderstorm came on in earnest, with sheets of rain, a regular *Wolkenbruch*.[66] The rain was most refreshing; but lightning is not a pleasant companion in presence of a bright ice-axe, and I was glad when the houses of Aviernoz came in sight. The village had the appearance of being lost; and the houses were scattered about so irregularly, that it was difficult to know which was the best point to make for. The road studiously avoided the scattered houses, and the *Mairie* seemed especially difficult to find. When at length it was found, the maire, like the queen in the poets, was in the kitchen; and he sat affably on the end of a bench and read the letter of introduction aloud, asking me, at the conclusion, how was our friend Dugravel, a man amazing in many ways. When I confessed that I had only made the acquaintance of the amazing man the night before, and therefore did not feel competent to give any reliable account of the state of his health, beyond the fact that he seemed to be in excellent spirits, the maire looked upon me evidently with great respect, as having won so far upon a great character like Dugravel in so short a time, and determined to accompany me himself. Meantime, we must drink some kirsch. The maire was a young man, spare and vehement. He talked with a headlong impetuosity which caused him to be always hot, and his hair limp and errant; and at the end of each sentence there were so many laggard halves of words to come out together, with so little breath to bring them out, that he eventuated in a stuttering scream. His clothes were of such a description, that the most speculative Israelite would not have gone beyond copper for his wardrobe, all standing. There were two women in the house, to whom he was exceedingly imperious: one of them received his orders and his vehemence with a certain amount of defiance, but the other was subdued and obedient, and I believe her to have been the mayoress. He poured himself and his household at my feet, knocked a child one way and his wife another, and, from the air with which he dragged off the tablecloth

---

[66] A Yorkshire farmer unconsciously adapts the German *Wolkenbruch*, declaring on occasion that the rain is so heavy, it is 'ommust as if a clood had brussen someweers.'

they had laid, and ordered a better, and swept away the glasses because they were not clean enough--which in itself was sufficiently true,--and screamed for poached eggs for monsieur, and then impetuously ate them himself--I fancy that he might have been taught to play Petrucio with success.

When we had sat for a quarter of an hour or so, a heavy-looking young man, in fustian clothes and last year's linen, came into the room, and was introduced as the communal schoolmaster. We shook hands with much impressment on the strength of the similarity of our professions, and the maire explained that the new arrival acted also as his secretary, for there was really so much writing to be done that it was beyond his own powers; and as the schoolmaster lived *en pension* at the *Mairie*, it was very convenient. M. Rosset, the schoolmaster, stated that he had heard us, as he sat in his room, talking of the proposed visit to the glacière, and he should much wish to accompany us. We both expressed the warmest satisfaction; but the maire suggested--how about the boys? That, M. Rosset said, was simple enough. The world would go to the school at nine o'clock, and, finding no schoolmaster, would go home again, or otherwise employ itself; and he could have school on the weekly holiday, to make up for the lost day. This weekly holiday is universally on Thursday, he said, because that day divides the week so well; and I failed to persuade him that there was a commemoration intended in the choice of that day, as in the observance of Friday and Sunday. The maire utterly refused to take a cord, on the ground that there was no possibility of such a thing being of the least use. Fortunately, I had now my own axe, which in more able hands had mounted more than once Mont Blanc and Monte Rosa, so I had not the usual fight to procure that instrument.

Half an hour from the *Mairie*, when we had well commenced the steep ascent of the mountain-side, the maire turned suddenly round and exclaimed, 'But the inspector!' Rosset was a sallow man, but he contrived to turn white, while M. Métral (the maire) explained to me that the inspector of schools was to visit Aviernoz that day. The schoolmaster recovered before long, and said he should inform the inspector that a famous *savant* had come from England, and required that the maire and the *instituteur* should accompany him to the glacière, to aid him in making scientific observations. In order that he might have documentary proof to advance, he asked for my card, and made me write on it my college and university in full.

As I have already said, the maire's style of talking required a good deal of breath, and so it was not unnatural that the ascent should reduce him to silence. The schoolmaster talked freely about scholastic affairs, and gave me an account of the ordinary tariff in village schools, though each commune may alter the prices of its school if it please. Under seven years of age, children pay 4 francs a year, or, for shorter periods than a year, at the rate of 75 centimes a month; between seven and thirteen, 6 francs a year, or 1 franc a month; from thirteen to eighteen, 8 francs a year, or 1 f. 50 c. a month. There is the same difficulty in France, of course, as with us, in keeping children at school after they are old enough to earn a few centimes by cattle-keeping; and the Ministry of Education had shortly before addressed questions to every schoolmaster in the country, asking what remedy each could suggest. My present friend had replied, that if the Government would give the education gratis, something might be done; but he had expressed his opinion that nothing short of an actual subsidy to parents of children beyond eight or nine years of age would ensure a general improvement.

Having given me this information, he observed that it was every man's business to learn, though he and I might be teachers also, and therefore he was sure monsieur would pardon him if he asked what those black patches on monsieur's hands might mean,--pointing to certain large areas of Epsom plaster which covered the tokens of many glacières. When his mind was set at rest as to this phenomenon, the maire called a halt, and took his turn of talking. He began to tell me about himself and his wealth, Rosset backing him up and putting in the most telling parts. He had very extensive property, and the more level parts of it were certainly valuable, consisting of 200 *journaux* of good arable land: the forests through which we walked were his, and he possessed three *montagnes* and châlets higher up on the mountain. The glacière was his own property; and two years ago he had discovered another in the neighbourhood, which he had not since visited.

He was assisted in his capacity of maire by twelve councillors--in a larger commune it would have been fifteen--and the council met four times in the year. If it was desirable that they should meet on any other occasion, he must write to the prefect of the arrondissement for permission, specifying the business which they wished to conduct, and to this specified business they must confine themselves entirely. Then he wished to know, had we maires such as he in England? Hereupon I drew a fancy picture of the Lord Mayor of London, receiving the Queen and the Royal Family in general in a friendly way, and giving them a dinner,--which, he observed, must cost a good deal, a great deal. However, he looked round upon his fields and houses and mountains, and seemed to think that he could himself stand a considerable drain upon his purse for the reception of royalty; and possibly he is now anxious that the Emperor should pass that way, during the five years to which the tenure of the mayoralty is restricted. Both of my companions were strong in their French sympathies--the one because under the new rule all communal affairs were so much better organised, the other because a wonderful change for the better had taken place in the government superintendence of schools. Theirs was formerly an odd corner of a kingdom that did not care much about them, and was not homogeneous; it was now an integral part of a well-ordered empire. They confessed that the present state of things cost them much more in taxes, &c., excepting in the upper mountains, where Rosset had a cousin who paid even less than under Sardinian rule.

Of course, we talked a little on Church questions; and they were astonished to hear that I was not only an ecclesiastic, but an ordained priest,--a sort of thing which they had fancied did not exist in the English Church. Rosset said the *curés* of small communes had about £40 a year, but I must have more than that, or I could not afford to travel so far from home. Had I already said the mass that morning? Had I my robes in the *sac* I had left at the *Mairie*? Was the red book they had seen in my hands (Bädeker's *Schweiz*) a Breviary? They branched off to matters of doctrine, and discussed them warmly; but some things they so accommodatingly understated, and others they stated so fairly, that I was able to tell them they were excellent Anglicans.

Higher up in the forest, we were nearly overwhelmed by a party of charcoal-porters, who came down with their *traîneaux* like a black avalanche. A *traîneau* is nothing more than a wooden sledge, on two runners, which are turned up in front, to the height of a yard, to keep the cargo in its place. In the more level parts the porter is obliged to drag this, but on the steep zigzags its own weight is sufficient to send it down; and here the porter places himself in front, with his back leaning against the sacks of charcoal and the turned-up runners, and the whole mass descends headlong, the man's legs going at a wild pace, and now one foot, now the other, steering a judicious course at the turns of the zigzags. The charcoal is made by Italians, who live on polenta and cheese high up in the mountains, and bring their manufacture down to a certain distance, after which the porters take it in charge. The men we saw told us that by hard work they could make four journeys in the day, earning a franc by each; out of which, as they said, they must support stomach and boots, one journey making them ready for a meal, and eight journeys finishing a pair of soles.

It cost us an hour and a half to reach the maire's first châlet, where we were to lunch on such food as the old woman who managed it might have on hand; that is to say, possibly bread, and, beyond that, milk only, in some shape or other. The forms under which milk can be taught to appear are manifold. A young Swiss student, who in the madness of his passion for beetle-hunting had spent fifteen days in a small châlet at Anzeindaz, sleeping each night on the hay,[67] gave me, some time since, a list of the various foods on which he lived and grew fat. The following is the *carte*, as he arranged it:--

Viandes.	Vins.
Du séret.	Du lait de vache.[68]
Du caillé.	Du lait froid.
Du beurre.	Du lait de chèvre.

[67] I tried the hay in this châlet one night, with such results that the next time I slept there, two years after, I preferred a combination of planks.

[68] i.e., New milk, warm.

Du fromage gras.	Petit lait.
Du fromage mi-gras.	De la crême.
Du fromage maigre.	Du lait de beurre.[69]
Tome de vache.	Petit lait de chèvre.
Tome de chèvre.	

Pour les Cochons.
Du lait gâté.
Cuite.

Some of the solids and fluids in the earlier part of this *carte* we felt tolerably sure of finding at the maire's châlet, and accordingly any amount of cream and *séret* proved to be forthcoming. The maire asserted that *cérac* was the true name of this recommendable article of food, *céré* being the patois for the original word. Others had told us that the real word was *serré*, meaning *compressed* curds; but the French writers who treat learnedly of cheese-making in the *Annales de Chimie* adopt the form *sérets*; and in the *Annales Scientifiques de l'Auvergne* I find both *seret* and *serai*, from the Latin *serum.*There was also bread, which arrived when we were sitting down to our meal: it had been baked in a huge ring, for convenience of carriage, and was brought up from the low-lands on a stick across a boy's shoulder. When the old woman thought it safe to expose a greater dainty to our attacks, at a later period of the meal, she brought out a pot of *caillé*, a delightful luxury which prevails in the form of nuggets of various size floating in sour whey. Owing to a general want of table apparatus, we placed the pot of caillé on a broken wall, and speared the nuggets with our pocket-knives.

After the meal, the two Frenchmen found themselves wet and exceedingly cold; for Frenchmen have not yet learned the blessing of flannel shirts under a broiling sun. They set to work to dry themselves after an original fashion. The fire was little more than a collection of smouldering embers, confined within three stone walls about a foot high;

[69] Otherwise graphically called *battu*.

so they took each a one-legged stool--*chaises des vaches*, or *chaise des montagnes*--and attached themselves to the stools by the usual leathern bands round the hips; then they cautiously planted the prods of the stools in the middle of the embers, maintaining an unstable equilibrium by resting their own legs on the top of the walls. Here they sat, smoking and being smoked, till they were dry and warm. Of course, in case of a slip or an inadvertent movement, they would have gone sprawling into the fire. A well-known Swiss botanist, who has seen many strange sleeping-places in the course of sixty years of flower-hunting in the mountains of Vaud and Valais, has told me that on one occasion he had reached with great difficulty the only châlet in the neighbourhood of his day's researches, at a late hour of the night, the whole mountain being soaked with rain. It was a little upland châlet, which the people had deserted for the autumn and winter; and meantime a mud avalanche had taken possession, and covered the floor to a depth of several inches. No plank was to be found for lying on; but he discovered a broken one-legged stool, and on this he sat and slept, propped as well as might be in a corner. It is difficult to say which would be worse--a fall from the stool by daylight into the embers of a wood fire, or the shuddering slimy waking about midnight, after a nod more vigorous than the rest, to find oneself plunged in eight cold inches of soft mud.

About half an hour beyond the châlet, we found the mouth of the glacière, on a large plateau almost bare of vegetation, and showing the live rock at the surface. They told me that in a strong winter there would be an average of 12 feet of snow on the ground here.[70] The glacière itself is approached by descending one side of a deep pit, whose circumference is larger than that of any other of the pit-glacières I have seen. A few yards off there is a smaller shaft in the rock, which we afterwards found to communicate with the glacière. The NW. side of the larger pit, being the side at the bottom of which is the arch of entrance, is vertical, and we spent the time necessary for growing cool in measuring the height of this face of

[70] I had no means of determining the elevation of the ground. The fact of 12 feet of snow is of no value as a guide to the height. Last winter (1864-5) there was 26 feet of snow on the Jura, at a height of less than 4,000 feet, and the position of some of the larger châlets was only marked by a slight boss on the plane surface.

rock from above. The plummet ran out 115 feet of string, and struck the slope of snow, down which the descent to the cave must be made, about 6 feet above the junction of the snow with the floor of the glacière, which was visible from the S. side of the edge of the pit; so that the total depth from the surface of the rock to the ice-floor was 121 feet.

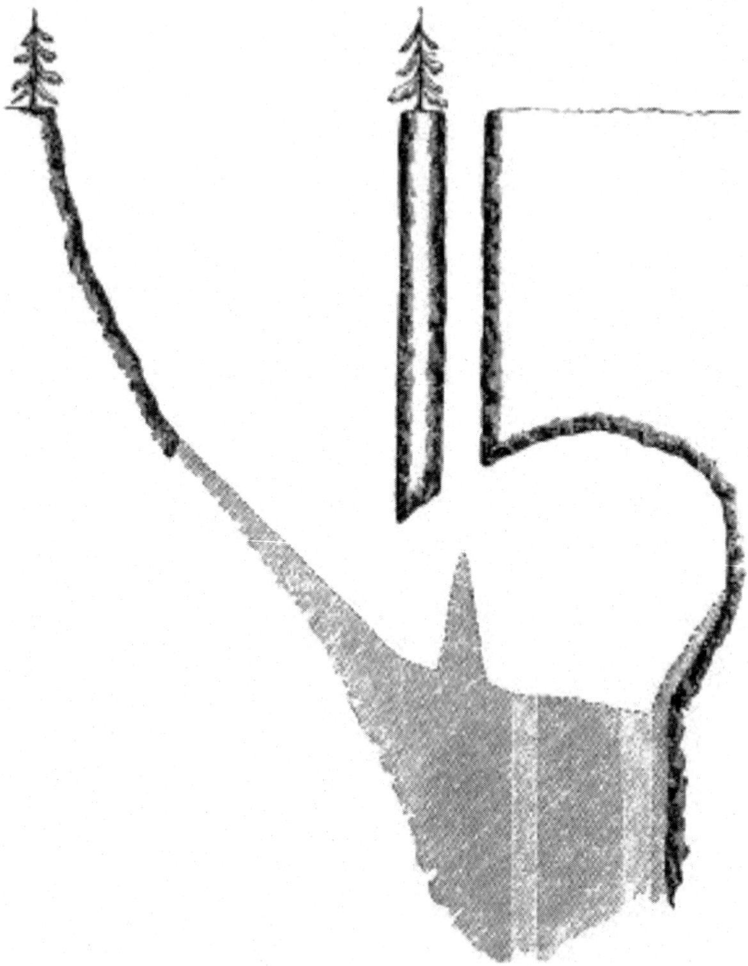

Figure 12. Vertical Section of the Glacière of Grand Anu, Near Annecy.

When we were sufficiently cool, we scrambled down the side of the pit opposite to that in which the archway lies, finding the rock extremely steep, and then came to a slope of 72 feet of snow, completely exposed to the weather, which landed us at the mouth of the glacière. The arch is so large, that we could detect the change of light in the cave, caused by the passage of clouds across the sun, and candles were not necessary, excepting in the pits shortly to be described. We saw at once that rapid thaw was going on somewhere or other; and when we stepped off the snow, we found ourselves in a couple of inches of soft green vegetable mud, like a *compote* of dark-coloured duckweed--or, to use a more familiar simile, like a mass of overboiled and ill-strained spinach. To the grief of one of us, there was ice under this, of most persuasive slipperiness. The maire said that he had never seen these signs of thaw in his visits in previous years; and as we went farther and farther into the cave, he was more and more surprised at each step to find such a large quantity of running water, and so much less ice than he had expected. The shape of the glacière is a rough circle, 60 feet in diameter; and the floor, which is solid ice, slopes gradually down to the farther end. The immediate entrance is half-closed by a steep and very regular cone of snow, lying vertically under the small shaft we had seen in the rock above. The snow which forms the cone descends in winter by this shaft; and the formation must have been going on for a considerable time, since the lower part of the cone has become solid ice, under the combined influences of pressure and of *dégel* and *regel.* I climbed up the side of this, by cutting steps in the lower part, and digging feet and hands deep into the snow higher up; and I found the length of the side to be 30 feet. I had no means of determining the height of the cave, and a guess might not be of much value.

At first sight, the farther end of the cave was the most striking. The water which comes from the melting snow down which we had passed in reaching the glacière, had cut itself deep channels in the floor, and through these it coursed rapidly till it precipitated itself into a large pit or *moulin* in the ice, at the lowest point. This pit, as will be seen by the section of the cave given on p. 174,[71] terminates the glacière; and the rock-wall at the

[71] In the section of the cave, I have brought out the deeper pit from the side into the middle, so as to show both in one section: I have also slightly shaded the pits, instead of leaving them blank like shafts in the rock.

farther edge falls away into a sort of open fissure, down which magnificent cascades of ice stream emulously, clothing that side of the pit, which would otherwise be solid rock. We cut a few steps about the upper edge of this *moulin*, to make all safe, and proceeded to let down a lighted candle, which descended safely for 36 feet, showing nothing but ice on all sides; it then came in contact with one of the falls of water, and the light was of course extinguished. We next tied a stone to the string, and found that after 40 feet it struck on ice and turned inwards, under our feet, stopping finally at the end of 51 feet; but whether it was really the bottom of the pit that stopped it, or only some ledge or accidental impediment, we could not determine. The diameter of this pit might be 3 yards, but we took no measure of it.

At the extreme right of the cave we found another pit, a yard and a half across, two-thirds of the circumference of which was formed by the plateau of ice on which we stood, and the remaining third by a fluting in the wall of rock. The maire said that, two years ago, this hole was not visible, being concealed by a large ice-column which had since fallen in. Here again I let down a lighted candle, with more hopes of getting it to the bottom, as no part of the cave drained into the pit. The candle descended steadily, the flame showing no signs of atmospheric disturbance, and revealing the fact that the opposite side of the pit, viz. the rock, which alone was visible from our position, became more and more thickly covered with ice, of exquisite clearness, and varied and most graceful forms. As foot after foot, and yard after yard, ran out, and our heads craned farther and farther over the edge of the pit to follow the descending light, (we lay flat on the ice, for more safety,) the cries of the schoolmaster became mere howls, and the maire lapsed into oaths heavy enough to break in the ice. It is always sufficiently disagreeable to hear men swear; but in situations which have anything impressive, either of danger or of grandeur, it becomes more than ever unbearable. I remember on one occasion over-taking a large party in the descent from the Plateau to the Grands Mulets, in a place where the snow was extremely soft, and any moment might land one of us in a crevasse; and I shall never forget the oaths which caught my ear, from a floundering fellow-countryman enveloped from the waist downwards.

When 60 feet had run out, the candle stopped, and on stretching over I saw that it had reached a slope of ice which inclined very steeply northwards, and passed away under the rock, apparently into a fresh cavern. By raising the candle slightly and then letting it drop, we made it glide down this slope for 8 feet; and then it finally rested on a shelf of ice, showing us the shadowy beginnings of what should be a most glorious ice-cave. The little light which the candle gave was made the most of by the reflecting material which surrounded it; and we were able to see that the archway in the rock was rounded off with grey ice, and rested, as it were, on icy pillars. As far as we could judge, there would have been abundant room to pass down the slope under the archway, if only the preliminary 60 feet could by any means have been accomplished; and I shall dream for long of what there must be down there.

As I was anxious to know whether the side of the pit was vertical ice under our feet, I contrived to get about a third of the way round the edge, so as almost to reach the fluting in the rock which formed the farther side of the pit, and then desired the schoolmaster to raise the candle slowly from the ledge on which it still rested. As he pulled it gradually up, I was startled to find that the ice fell away sharply immediately below the spot where we had been collected, and then formed a solid wall; so that we had been standing on the mere edge of a shelf, with nothing but black emptiness below. How far the solid wall receded at the bottom I was unable to determine, for the light of one candle was of very little use at so great a distance, and in darkness so profound. I persuaded the maire to make an effort to reach a point from which he could see the insecurity of the ice which had seemed to form so solid a floor; and he was so much impressed by what he saw, that he fled with precipitation from the cave, and we eventually found him asleep under a bush on the rocks above. In reaching the farther side of the pit, we crossed unwittingly an ice-bridge formed by a transverse pit or tunnel in the ice, which opened into the pit we were examining. The maire afterwards promised to rail off all that end of the glacière, and forbid his workmen to venture upon it. Considering that the hole itself was only opened two years before by the fall of a column, and has already undergone such changes, I shall be surprised if the ice-bridge, and all that part on which we lay to fathom the pit, does not fall in before very long; and then, by means of steps and ropes and ladders, it may be possible to reach the entrance to the lower cave, 190 feet below the surface of the earth. May I be there to see![72]

The left side of the glacière, near the entrance, was occupied by a columnar cascade, behind which I forced a passage by chopping away some lovely ornaments of ice. Here also the solid ground-ice falls away a little under the surface, leaving a cavern 8 or 9 feet deep, on the rock side of which every possible glacial fantasy was to be found. The stalactites here presented the peculiar prismatic structure so often noticed; but on the more exposed side of the column they were tipped with limpid ice, free from all apparent external or internal lines. This reminded me of what we had observed in the Glacière of La Genollière, namely, that the surface-lines tended to disappear under thaw; so I cut a piece of prismatic ice and put it in my mouth. In a short time it became perfectly limpid, and on breaking it up I could discover no signs of prism. On some parts of the floor of the glacière, the ice was apparently unprismatic, generally in connection with running water or other marks of thaw; but, to my surprise, I found that it split into prisms very readily.

The maire could not understand how it was that, after a winter especially severe, as that of 1863-4 had been, there should be even less ice than in the preceding summer, and we could see the marks of last year's cutting, down to the edge of the *moulin*. He said that they had never before cut down in that direction; but in the summer of 1863 they had been so much struck by the clearness of the ice which formed the floor, that they had cut it freely, and removed a large quantity. This, I believe, was the cause of the absence of any great amount of fresh ice. The slope of the whole ice-floor is considerable, and the workmen increased the slope by cutting away the ice in the neighbourhood of the edge of the *moulin*: they had also, as we could see quite plainly, excavated the clearer parts of the ice between the entrance to the cave and the *moulin*, so that a sort of trough ran down from near the foot of the snow to the pit at the lower end of the

[72] I have made arrangements for completing the exploration of this cave, and the one which is next described, in the course of the present summer.

glacière. When we were there, the water rushed down this trough, and was lost in the pit; and very probably the same may have been the case in the earlier parts of the year, when, according to the view I have already expressed, the ice would under ordinary circumstances have been formed. If this be so, the caverns below must have received immense additions to their stores of ice or water. We observed, by the way, that the slope of ice to which the candle descended in the deeper pit, and the shelf on which it rested, were quite dry, or at any rate free from all apparent signs of the abundant water we should have seen, had that been the outlet for the streams which poured into the *moulin*. The maire said that the columns and cascades of ice in the cave had been much more beautiful in the previous summer.

The whole cavern would thus appear to be something of the shape of an egg, with the longer axis vertical, and the entrance about half-way up the side. The lower end of this egg-shaped cavity in the rock is filled with ice, which in some parts shrinks from the rock below the surface, though, as far as outward appearance goes, it fills the cavern to its farthest corners. The depth of this ice at one side is 60 feet, and how much more it may be in the middle it is impossible to say. As we have seen, there is a second ice-cave opening out of the principal one, at a depth of 190 feet below the surface; and with respect to this second cave imagination may run riot. Rosset told me that he had noticed, the year before, a strong source of water springing out of the side of a rock, at some little distance from the glacière; but he could not reach it then, and could not find it now. This may possibly be the drainage of the glacière in its summer state.

The thermometer stood at 34° in the middle of the cave; and though the others felt the cold very much, I was myself surprised to find so low a register, for the atmosphere seemed to be comparatively warm, judging from what I had experienced in other glacières. The only current of air we could detect was exceedingly slight, and came from the deeper of the two pits in the ice. It was so slight, that the flame of the candle burned apparently quite steadily when we were engaged in determining the depth and shape of the pit.

The sun had by this time produced such an effect upon the slope of snow outside the glacière, that we found the ascent sufficiently difficult, especially as our hands were full of various instruments. The schoolmaster was not content to choose the straight line up, and in attempting to perform a zigzag, he came to a part of the slope where the snow lay about 2 inches thick on solid ice, and the result was an unscholastic descent in inverted order of precedence. He got on better over the rolling stones after the snow was accomplished, but the clumsy style of his climbing dislodged an unpleasant amount and weight of missiles; and though he was amiable enough to cry '*Garde*!' with every step he took, it will be found by experiment that it is not much use to the lower man to have '*Garde*!' shouted in his ears, when his footing is insecure to begin with, and a large stone comes full at his head, at the precise moment when two others are taking him in the pit of the stomach.

We found the maire, as was said, asleep under a bush near the mouth of the pit; and he pronounced himself completely recovered from the effects of the cold, and ready to guide us to a second glacière. He told us that the amount of ice he sold averaged 4,000 *quintaux métriques* a week, for the three months of July, August, and September; but the last winter had been so severe, that the lake had provided ice for the artificial glacières of Annecy, and no one had as yet applied to him this year. As only a fortnight of his usual season had passed, he may have since had plenty of applications, later in the year. The railways have opened up more convenient sources of ice for Lyons, and for some time he has sent none to that town.

Chapter 11

The Glacière of Chappet-Sur-Villaz, on the Mont Parmelan, Near Annecy

We started southwards from the Glacière of *Grand Anu*, for such they said was the proper name for the cave last described, and passed over some of the wildest walking I have seen. All the most striking features of a glacier were here reproduced in stone: now narrow deep crevasses which only required a slight spring; now much more formidable rents, which we were obliged to circumvent by a détour; now dark mysterious holes with vertical shell-like partitions at various depths; and now a perfect *moulin*, with fluted sides and every detail appertaining to those remarkable pits, the hollow plunge of falling water alone excepted. In other parts, the smooth slab-like appearance of the surface reminded me of a curious district on one of the summits of the Jura, where the French frontier takes the line of crest, and the old stones marked with the *fleur-de-lys* and the Helvetic cross are still to be found. In those border regions the old historic distinctions are still remembered, and the frontier Vaudois call the neighbouring French *Bourguignons*--or, in their patois, *Borgognons*. They keep up the tradition of old hatreds; and the strange bleak summit, with its smooth slabs of Jura-chalk lying level with the surface, is so much like a vast cemetery, that the wish in old times has been father to the thought, and they call it still the Cemetery of the Burgundians, *Cimetiros ai Borgognons*.[73]

After a time, we reached a tumbled chaos of rock, much resembling the ice-fall of a glacier, and, on descending, and rounding a low spur of the mountain so as to take a north-westerly course, we found ourselves in a perfect paradise of flowers. One orchis I shall always regret. There seemed to be only a single head, closely packed with flowerets, and strongly scented; it was a pure white, not the green and straw-coloured white of other scented orchises. There were large patches of the delicate *faux-lis (Paradisia liliastrum)*; and though there might not be anything very rare, and the lovely glacier-flowers were of course wanting, the whole was a rich feast for anyone who cares more for delicacy and colour than for botany.

The maire told us that he had found the glacière, for which we were now in search, two years before, when he accompanied the government surveyor to show him the forests and mountains which formed his property. As he had on that occasion approached the spot from the other side, we walked a long way to place him exactly where the surveyor and he had crossed the ridge of the mountain, and then started him down from the Col in the direction they had taken. He was certain of two things: first, that they had passed by the Col between the Mont Parmelan and the Montagne de l'Eau; and, secondly, that the glacière was within five minutes of the highest point of the Col. For three-quarters of an hour we all broke our shins, and the officials the Third Commandment. They invoked more saints than I had ever heard of, and, in default, did not scruple to appeal with shocking volubility to darker aid. It was all of no use,--and well it might be; for when we had given it up in despair, after long patience and a considerable period of the contrary, and had descended for half an hour in the direction of a third glacière, I chanced to look back, and saw that the Col in the neighbourhood of which we had been searching lay between two points of the Montagne de l'Eau; while the true Col between that mountain

---

[73] The true *Cimetière des Bourguignons* is the enclosure where René, the victor of Nancy, buried the Burgundians who fell on the sad Sunday when Charles the Bold went down before the deaf châtelain Claude de Bagemont.

and the Mont Parmelan lay considerably to the west. When it appears that a guide has probably made a mistake, the only plan is to assume quietly that it is so, as if it were a matter of no consequence, and then he may sometimes be decoyed into allowing the fact: I therefore pointed out to the maire the true Col, and told him that was the one by which he had passed southwards, when he found the glacière; to which, with unnecessary strength of language, he at once assented. But all my efforts to take him back were unavailing. Nothing in the world should carry him up the mountain again, now that he had happily got so far down. I worked his best and his worst feelings with equal want of success; even national jealousy failed, and he was content to know that a French maire had not pluck to face three-quarters of an hour of climbing, when an English priest was ready to lead the way. The schoolmaster declined to go alone with me, on the ground that neither of us knew the mountain, and threatening clouds were gathering all around. When, at last, I proposed to go by myself, they became menacingly obstructive, and declared that I should certainly not be allowed to face the intricacy of the mountain in a fog. Besides, as the maire put it, he was sure of the way to the third glacière; and if I were to go up alone to look for the second, I should lose a certainty for a chance, as there was not time to visit both. So with an ill grace I continued the descent with them, being restored to good humour before long by the beauty of the Lake of Annecy, as seen from our elevated position.

It is so impossible to accept in full the accounts one picks up of natural curiosities, that I give the maire's description of the stray glacière only for what it is worth. It was not extracted without much laborious cross-examination--*sais paw vous le dire* being the average answer to my questions. The entrance to the cave is about twice as high as a man, and is in a small shallow basin of rock and grass. The floor is level with the entrance, and the roof rises inside to a good height. In shape it is like a Continental bread-oven; and at the time of the maire's visit, the floor was a confused mass of ice and stones, the former commencing at the very entrance. There was no ice except on the floor, the area of which might be as large as that of the surface of the ice in the Glacière of Grand Anu. No pit was to be seen, and not a drop of water. Snow could have drifted in easily, but they saw no signs of any remaining. If this account be true, especially with respect to the position of the entrance and the horizontal direction of the floor, I have seen no glacière like it.

We descended for a time through fir-woods, and then again down steep and barren rocks, till we reached the sharp slope of grass which so frequently connects the base of a mountain with the more civilised forests and the pasturages below. The maire led us for some distance along the top of this grass slope, towards the west, skirting the rocks till they became precipitous and lofty, when he said we must be near our point. Still we went on and on without seeing any signs of it, and our guide seemed in despair; and I, for one, entirely gave up the third cave to the same fate as the second, and became very sulky and remonstrative. The entrance to the glacière, the maire told us, was a hole in the face of the highest rocks, 3 or 4 yards only above the grass; and as we had now reached a part of the mountain where the rock springs up smooth and high, and we could command the whole face, and yet saw nothing, the schoolmaster came over to my side, and told the maire he was a humbug. However, we were then within a few yards of the desired spot, and half-a-dozen steps showed us a small *cheminée*, down which a strong and icy current of wind blew. The maire shouted a shout of triumph, and climbed the *cheminée*; and when we also had done the necessary gymnastics, we found a hole facing almost due north, all within being dark. The current blew so determinedly, that matches were of no use, and I was obliged to seek a sheltered corner before I could light a candle; and, when lighted, the candle was with difficulty kept from being blown out. No ice was visible, nor any signs of such a thing,--nothing but a very irregular narrow cave, with darkness at the farther end. As we advanced, we found that the floor of the cave came to a sudden end, and the darkness developed into a strange narrow fissure, which reached out of sight upwards, and out of sight below; and down this the maire rolled stones, saying that *there* was the glacière, if only one could get at it without a *tourneau*. Considering the persistency with which he had throughout declared that there was no possible need for a rope, I gave him some of my mind here, in that softened style which his official dignity demanded; but he excused himself by saying that the gentleman who owned the glacière, and extracted the ice for private use only, was now living at his summer châlet, a mile or two off, and he, the maire, had felt confident that the *tourneau* would have been fitted up for the season.

On letting a candle down from the termination of the floor, we found that the perpendicular drop was not more than 12 feet, and from the shelf thus reached it seemed very possible to descend to the farther depths of the fissure; but I had become so sceptical, that I persisted in asserting that there was no ice below. The maire's manner, also, was strange, and I suspected that the cold current of air had caused the place to be called a glacière, with any other qualification on the part of the cave. One thing was evident,--no snow could reach the fissure. M. Métrai was determined that I must not attempt the descent, pointing out, what was quite true, that though the fall was not great, there seemed no possibility of getting back up the smooth rock. His arguments increased my suspicions; so, leaving all apparatus behind, I dropped down to join the candle, rather hoping to have the satisfaction of sending them off for a rope, in case I could not achieve the last few feet in returning, and knowing that there was no danger of the fate which once threatened the chamois-hunting Kaiser Max.[74]

The drop turned out to be a mere nothing, and, taking the candle, I scrambled on, down the sloping floor of the fissure, towards the heart of the mountain, expecting every moment that my further passage would be stopped by solid rock. But, after reaching a part so narrow that I was obliged to mount by both sides at once in order to get past it, I found a commodious gallery, opening out into a long and narrow and very lofty cavern, still only a fissure, the floor of which continued the regular and rapid slope down which I had so far come. A short way farther down, an opening appeared to the left; and I turned off the main passage into a horizontal gallery or chamber, with a floor of ice resting on rock and stones. This chamber seemed to be 3 or 4 yards wide at the entrance, narrowing regularly to 4 1/2 feet. It was 40 feet long, and at the farther end, which would not have been visible from the entrance, on account of a slight bend in the ice-gallery, even if there had been any light, it was closed

---

[74] Neither of my companions, I fear, would have acted as Sejanus did, when another emperor was in danger of his life in the cave on the Gulf of Amyclæ. (Tacit. Ann. iv. 59.)

by an ice-cascade 7 yards high and 4 1/2 feet broad at the bottom. The ice of much of this cascade was so clear, that I saw the rock upon which it rested, or in some parts did not rest, quite plainly, and the large air-cavities in the structure were beautifully shown by the richly-coloured rock behind. None of the current which we had observed above, and which had nearly baffled my protecting care of the candle during the descent, came from this gallery; but I find it written in my notes that the gallery was *very* cold. Thaw was going on, rather rapidly; and the water stole out by the entrance, and ran down the main descent, over ice and among rocks, into the farther darkness.

When I came out again from this gallery, I mounted the slope towards my companions, and tried to tempt them down. The maire felt himself to be too valuable to his country to be lightly risked, and declined to come; but Rosset took a bold heart, and dropped, after requiring from me a solemn promise that I would give him a back for his return up the rock. We visited the gallery I had already explored, and, as we stood admiring the cascade of ice, a skilful drop of water came from somewhere, and extinguished our only candle. My matches were with the maire; and I was equally sure that he would not bring them down to us, and that we could not go up to fetch them without a light. Rosset, however, very fortunately, had a box in his pocket for smoking purposes; and we cut off the wet wick, and cut down the composition to form another, and so contrived to light the candle again. While we were thus engaged, I chanced to look up for a moment, and saw far above our heads a small opening in the roof, through which a few rays of light entered from the outer world. It was so very far above us, that the uncertain rays were lost long before they got down to our level, being absorbed in the universal darkness, and being in fact rather suggested than visible even at their strongest. Those who have been at Lauterbrunnen in a very dry season, will understand how these rays presented the appearance of a ghostly Staubbach of unreal light. We must have been at an immense depth below the surface in which the opening lay; and if there had been a long day before us, it would have been curious to search for the fissure above. Sir Thomas Browne says, in the *Religio*

Medici, 'Conceive light invisible, and that is a spirit.' We very nearly saw a spirit here.

The descent from the mouth of this chamber to the deeper recesses of the main fissure was very rough, but was speedily accomplished, and we reached a point where solid rock stopped us in face; while, to the right, a chamber with a threshold of ice was visible, and, to the left, a dark opening, down which the descent appeared to continue. From this opening all the strong cold current came. We took the ice-chamber first.

The entrance had evidently been closed till very lately by a large column of ice, and we passed over the débris, between rock portals and on a floor of solid grey ice, into a triangular cave of any height the imagination might choose to fix. The entire floor of the cave was of ice, giving the impression of infinite thickness and firmness. A little water stood on it, near the threshold, so limpid that we could not see where it commenced. The base of this triangular floor we found to be 17 feet, and its altitude 30 feet; and though these dimensions may seem comparatively small, the whole effect of the thick mass of ice on which we stood, with the cascades of ice in the corners, and the ice-figures on the walls, and the three sides of the cave passing up into sheer darkness, was exceedingly striking, situated, as it all was, so deep down in the bowels of the earth. The original entrance to the fissure, at the top of the *cheminée*, was, as has been said, at the base of lofty rocks, and we had descended very considerably from the entrance; so that, even without the strange light thrown upon the matter by the small hole overhead, through which we had seen the day struggling to force its way into the cavern, we should have been sure that we were now at an immense distance below the surface. One corner of the cave was occupied by a broad and solid-looking cascade, while another corner showed the opening of a very narrow fissure, curved like one of the shell-shaped crevasses of a glacier. Into this fissure the ice-floor streamed; and Rosset held my coat-tails while I made a few steps down the stream, when the fall became too rapid for further voluntary progress. I let down a stone for 18 feet, when it stuck fast, and would move neither one way nor the other. The upper wall of this fissure was clothed with moss-like ice, and ice of the prismatic structure,--with here and there large scythe-blades, as it were, attached by the sharp edge to the rock, and lying vertically with the heel outwards. One of these was 11 inches deep, from the heel to the rock, and only one-eighth of an inch thick at the thickest part.

The angle occupied by the cascade or column was the most striking. The base of the column was large, and apparently solid, like a smooth unbroken waterfall suddenly frozen. It fitted into the angle of the cave, and completely filled up the space between the contiguous walls. I commenced to chop with my axe, and before long found that this ice was hollow, though very thick; and when a sufficient hole was made for me to get through, I saw that what had looked like a column was in truth only a curtain of ice hung across the angle of the cave. Within the curtain the ice-floor still went on, streaming down at last into a fissure something like that in the other corner. The curtain was so low, that I was obliged to sit on the ice inside to explore; and after a foot or two of progress, the slope towards the fissure became sufficiently great to require steps to be cut. The stream of ice turned round a bend in the fissure, very near the curtain, and was lost to view; but Rosset stood by the hole through which I had passed--on the safer side of it--and despatched blocks of ice, which glided past me round the corner, and went whizzing on for a long time, eventually landing upon stones, and sometimes, we fancied, in water. It is very awkward work, sitting on a gentle slope of the smoothest possible ice, with a candle in one hand, and an axe in the other, cutting each step in front; especially when there is nothing whatever to hold by, and the slope is sufficient to make it morally certain that in case of a slip all must go together. Of course, a rope would have made all safe. When I groaned over the maire's obstinacy, Rosset asked what could possibly be the use of a rope, if I were to slip; and, to my surprise, I found that he had no idea what I wanted a rope for. When he learned that, had there been one, he would have played a large part in the adventure, and that he might have had me dangling over an ice-fall out of sight round the corner, he added his groans to mine, and would evidently have enjoyed it all very much. At the same time, he was prudent, and, as each block of ice made its final plunge, he told me that was what would happen to me if I went any farther: and, really, the pictures he drew of deep lakes of icy water and jagged points of rock, between which I must make my choice down there, were so unpleasant, that at last I desisted, and pushed myself up backwards, still in a sitting posture, calling Rosset and the maire the worst names I could feel justified in using. On the way, I found one of the large brown flies which we had seen in the Glacière of La Genollière, and in the Lower Glacière of the Pré de S. Livres.

Rosset now told me he was so cold he could stand it no longer; but, after a little pressure, and a declaration on my part that he should not have a candle for going up again, he consented to remain with me while I explored the remaining chamber, the lowest of all. This chamber may be called a continuation of the main passage. It is of about the same width as the highest of the three chambers, and the floor descends rapidly, the cold current of air becoming very strong and biting as we penetrated into the darkness. As the Genevese *savans* seemed to believe in 'cold currents' as the cause of underground ice, I was naturally anxious to see as much as possible of the state of this gallery, from which every particle of the current seemed to come. We very soon reached a narrow dark lake, and, exclaiming that here was ice again, I stepped, not on to, but into it, and found that it was water. When our solitary candle was brought to bear upon it, we saw that it was so clear as not in any way to impede our view, producing rather the effect of slightly-clouded spectacles upon the stones at the bottom. This lake filled up the whole breadth of the gallery, here perhaps 4 or 5 feet, and rapidly passed to the depth of a yard; but for a little distance there were unstable stones at one edge, and steps in the rock-wall, by which I could pass on still into the darkness, supported by an alpenstock planted in the water. The current of cold air blew along the surface of the water from the farther extremity of the gallery, wherever that might be. As far as our eyes could reach, we saw nothing but the black channel of water, with its precipitous sides passing up beyond our sight. It might have been possible to progress in a spread-eagle fashion, with one hand and one foot on each side; but a fall would have been so bitterly unpleasant, that I made a show of condescension in acceding to Rosset's request that I would not attempt such a thing. In the course of my return to the rocks where he stood, I involuntarily fathomed the depth of the lake, luckily in a shallower part, and was so much struck by the coldness of the water, that I left Rosset with the candle, and struggled up without a light to the place where we had left the maire, or rather to the bottom of the drop from the entrance-cave, to get the thermometer. The maire was sunning himself on the rock, out of reach of the cold current; but he came in, and let down the case, and I quickly rejoined the schoolmaster. At first, it would have been impossible to move about without a light; but our eyes had now become to some extent accustomed to the darkness, and I had learned the difficulties of the way.

When the thermometers were suspended in the water, Rosset asked how long they must stay there. I rashly answered, a quarter of an hour; on which he demanded indignantly whether I supposed he meant to stay in that cold for a quarter of an hour. He had now the candle in his own possession, and I was propped on a stone and an alpenstock in the lake, so he turned to go, vowing that he would leave me alone in the dark if I did not come out at once. There was no help for it, as the thermometer would have been of no use without a candle, and a step in the dark is not pleasant when all around is water, so I slowly drew up the thermometer and read 33° F. In making final arrangements for departure, I let it lie in the water for a few seconds longer, and it fell to 32½°; but Rosset would not stay a moment longer, and I was obliged to be content with that result. He made himself very easy about the matter, and said we must call it zero; and in the evening I heard him telling the maire that the greatest of the wonders he had missed, by his patriotic care for his neck, was a lake of water which did not freeze, though its temperature was zero (centigrade).

Among the stones at the bottom of this water, I saw here and there patches of a furry sort of ice. I have often watched the freezing of a rapid Scotch stream, where, in the swifter parts, the ice forms first at the bottom and gradually creeps up the larger stones till it appears on the surface, and becomes a nucleus, round which pieces of floating ice collect; and the substance in the glacière-lake had exactly the same appearance as the Scotch ground-ice. But it could not be the same thing in reality, for, as far as I understand the phenomenon of ground-ice, some disturbed motion of the water is necessary, to drive down below the surface the cold particles of water, which become ice the moment they strike upon any solid substance shaped like fractured stone;[75] the specific gravity of freezing water being so much less than that of water at a somewhat higher temperature, that without some disturbing cause it would not sink to the bottom.[76] So that it seems probable that the ice at the bottom of the lake was the remains of a solid mass, of which the greater part had been converted into water by some warm influence or other. We noticed that a little water trickled down among the stones which formed the slope of descent into the lowest gallery, so that perhaps the lake was a collection of water from all parts of the various ramifications of the fissure. Whence came the icy wind, it is impossible to say, without further exploration. It was satisfactory to me to find that the 'cold current' of the Genevese *savans* was thus associated with water, and not with ice, in the only cave in which I had detected its presence to any appreciable extent, the currents of the Glacière of Monthézy being of a totally different description.

When we reached the final rock, in ascending, I offered Rosset the promised back, but he got up well enough without it. Before leaving the entrance-cave, we inspected the thermometer which we had left to test the temperature of the current of air, and, to my surprise, found it standing at 48°. We saw, however, that it had been carelessly propped on a piece of rock which sheltered it from the influence of the current, so I exposed it during the time occupied in arranging the bag of tapes, &c., and it fell to 36°: whether it would have fallen lower, the impatience of Rosset has left me unable to say. If I can ever make an opportunity for visiting the Mont Parmelan again, I shall hope to take a cord, in order to investigate the mysterious corner of the triangular chamber; and I shall certainly make myself independent of shivering Frenchmen while I measure the temperature of the lake and the current of air. We met a man outside who said that he was employed by the owner, M. de Chosal of Annecy, to cut the ice; he had been down three times to the lowest gallery in different

[75] Water reduced to a temperature below 32° without freezing, begins to freeze as soon as a crystal is dropped into it, the ice forming first on the faces of the crystal.

[76] Water attains its maximum of specific gravity at 40°. Below 40° it becomes lighter.

years, in the end of July, and had always found the same collection of water there. The glacière, he told us, was discovered about thirty years ago.

The maire had basked in the sun all the time we were down below, and he expressed himself as much pleased that we had found so much to interest us, in spite of the miscarriage of our efforts to reach the second glacière. We set off down the steep grass at a scrambling sliding run, against which I was speedily obliged to protest, explaining that a certain ugly inflammation above the left knee was becoming worse every other step, and as the leg must last three days longer, it would be as well to humour it. They saw the force of this reasoning, and we descended with much gravity till we came in sight of the *Mairie*, still half an hour off, when Rosset cried out that he smelled supper, and rushed off at an infectious pace down the remainder of the mountain-side.

We reached the *Mairie* at six o'clock, and sat down at once 'to eat something.' The first course was bread and kirsch; and when that was finished, six boiled eggs appeared, and a quart *carafe* of white wine. These having vanished, their place was taken by a dish of sodden cabbage, and another quart of wine; but, to save the credit of the maire and the schoolmaster, I will not say how often the former functionary descended to the cellar with a quart pitcher, with increasing impetuosity. Next came a dish of onions, with a pretence of *mange-tout,* broiled brown after boiling, and served in a compound fat; and then haricots with a like condiment, and with a flavour reminiscent of the previous course. There was some talk of a *poulet*; but the bird still lived, and the talk came to nothing. The dinner ended with the haricots, and we then relapsed into dessert, namely, bread and kirsch. The mayoress came in with the dessert, and sat on the end of the bench, below the hats and the bread-tin, eating the remaining onions off the dish with the spoon of nature.

During one of the maire's frequent visits to the cellar, I propounded a question to the schoolmaster which had puzzled me for some time: Was I to pay the maire? M. Rosset said that it was certainly not *necessary*, but I had better propose it, and I should then see how M. Métral took it. This I accordingly did, when the adieux in the house had been said, and my host was showing me the way to Thorens, where I was to sleep, he, also, declared that it was not necessary--the pleasure he had experienced in accompanying me had already fully recompensed him: still, if I wished to reimburse him for that which I had actually cost, he was a man reasonable, and in all cases content. I calculated that the dinner and wine which had fallen to my share would be dear at a franc, and the day's wage of a substitute to do the maire's neglected work could not come to much, so I boldly and unblushingly gave that great man four francs, and he said regretfully that it was more than enough. To his son and heir--the identical boy who had brought the ring of bread up the mountain to the chalet where we lunched. I gave something under two-pence, for guiding me across two doubtful fields into a beaten track, and he expressed himself as even more content than the maire. They both told me that it was impossible to miss the way; but I imagine that I achieved that impossibility, as I had to walk through two streams in the deepening twilight, and the prevailing fear of water in that region is very considerable.

The *auberge* at Thorens to which the maire had recommended me, as being the best, and kept by a personal friend of his, bore the sign *à la Parfaite Union*. The entry was by the kitchen, and through the steam and odour of onions, illuminated by one doubtful oil-lamp, I saw the guest-room filled with people in Sunday dress, while two fiddles played each its own tune in its own time. Nothing but the potent name of M. the Maire of Aviernoz gained me even a hearing; and, for a bed, I was obliged to stretch my intimacy with that exalted personage to the very furthest bounds of truth. Chappaz Nicolai, whose name the maire had written in my note-book, that there might be no mistake, appeared to be of that peculiar mental calibre which warrants Yorkshire peasants in describing a man as 'half-rocked,' or 'not plumb.' His wife, on the other hand, was one of those neat, gentle, sensible women, of whom one wonders how they ever came to marry such thick-lipped and blear-eyed men. Between them they informed me that if I did not object to share a room, I could be taken in; otherwise--maire or no maire--not. I asked whether they meant half a bed; but they said no, that would not be necessary at present; and I accepted the offered moiety of accommodation, as it was now seventeen hours since I had started in the morning, and I was not inclined to turn out in the dark to look for a whole room elsewhere.

The stairs were a sort of cross between a ladder and nothing, and when we reached the proposed room a large mastiff was in possession, who would not let us enter till the master was summoned to expel him. The furniture consisted of a table and five chairs, with no bed or beds. On the chairs were various articles of clothing, blouses and garments more profound, belonging probably to members of the party below; and on the table, a bottle of water and a soup-plate, the pitcher and basin of the house. It was a mere slip of a room, with two diamond-shaped holes in one wall, whose purpose I discovered when my guide opened a papered door, in which were the holes, and displayed two beds foot to foot in an alcove. One of these, she was sure, would be too short for me, but she feared I must be satisfied with it, as the other was much broader and would therefore hold the two messieurs. How the *two*? I asked, and was told that two *pensionnaires* lived in this room; but they were old friends, and for one night would sleep in the same bed to oblige monsieur. The ideas of length and breadth in connection with the beds were entirely driven from my head by the fact of their dirtiness; and I determined that if the two *pensionnaires* occupied the one, the other should be unoccupied.

After arranging things a little, I struggled down the steps again, and ordered coffee and bread in a little room, which commanded the assembly with the fiddles in the larger *salle*. The head waitress, busy as she was, found time to come now and then to an open window near where I sat, and talked to a male friend sitting outside in the dark: indeed, she did more than talk, and people had to rattle their glasses very hard before they could make her hear. From her I learned that this was a marriage party which had arrived; and when I asked why they did not dance, as the fiddles were engaged at that moment with unwonted unanimity upon dance-music, she gave me to understand that these were not people of Thorens, but only a party from another village, making the evening promenade after the wedding: from which it would seem that it is not the etiquette for people to dance under such circumstances, except in the home village. They sat round a table, men and women alternately, with their hats on, and with glasses before them. The bride and bridegroom were accommodated with a bench to themselves at the head of the table, he likewise with his hat on, and with a pipe in his mouth, which, seeing that he was a demonstrative bridegroom, one might have supposed to be an inconvenience. He managed very well, however, and everyone seemed contented: indeed, the pipe must, I think, be held to be no difficulty; for the men all smoked, and yet, to judge from appearances, there was a prospect of as many marriages as there were couples in the room. The unruffled gravity, however, and the apparent want of zest, both in giving and receiving, which characterised the proceedings specially referred to, led me to suppose that it might be only a part of the etiquette, and so meant nothing serious.

Between ten and eleven the fiddles and the party vanished, and I went up-stairs more determined than ever not to touch a bed, after my experience of the room below. Three chairs were speedily arranged between the table and wall, and on these I lay and tried to sleep. But the very chairs were populous, as I had found below, and sleep was impossible. Moreover, soon after eleven, a soldier came into the room, to arrange about his breakfast with one of the maidens in the house. He had heard me order fresh butter for six o'clock, and he was anxious to know, whether, by breakfasting at five o'clock, he could get my butter. The chairs which formed my bed were under the lee of the table, so that the figure recumbent on them was invisible, and the gallant soldier, under the impression that there was no one in the room, enforced his arguments by other than conventional means. But military lips, when applied personally, proved to be a rhetoric as unsuccessful as military words. The maid was platonic, and something more than platonic; and the hero got so much the worst of it, that he gave up the battle, and changed the subject to a conscript in his charge, who had locked himself in his bed-room and would not answer. How was he to know whether he had the conscript safe? All this lasted some time; and when they were gone, one of the *pensionnaires* came in. With him I had to fight the battle of the window, which I had opened to its farthest extent. After he had got over the first surprise and shock of finding me on the chairs instead of in the bed, for whose comfort he vouched enthusiastically, he became confident that it was merely out of complaisance to him and his comrade that I had opened the window, and assured me that they really did not care for fresh air, even if they could feel the difference in the alcove, which he declared they could not. As soon as that was arranged to my satisfaction, the other *pensionnaire* came in, and with him the battle was fought with only half success, for he peremptorily closed one side of the window. He was a particularly noisy *pensionnaire*, and shied his boots into every corner of the room before they were posed to his satisfaction. As far as I could tell, the removal of the boots was the only washing and undressing either of them did; and then they arranged their candles in the alcove, lighted cigars, and got into bed. There the wretches sat up on end, smoking and talking vehemently, till sheer exhaustion came to my aid, and I fell asleep; but the edges of the rush-bottomed chairs speedily became so sharp that a recumbent posture ceased to be possible, and I sat dozing on one chair. A little before four o'clock, the noisier man got up to look for his boots; and as the friends continued their discussion, I also turned out and made for the nearest stream, where I bathed in a rapid at half-past four, to wash away, if possible, the horrors of the night.

Chapter 12

The Glacières of the Brezon, and the Valley of Reposoir

The bill *à la Parfaite Union* was as small as the accommodation at that *auberge*, and it was an immense relief to get away from the scene of my sufferings. The path to Bonneville lies for the earlier part of the way through pleasant scenery; and when the highest ground is reached, there is a lovely view of the Lake of Geneva, which may be enjoyed under the cool shade of a high hedge of trees, in the intervals of browsing upon wild strawberries. But after passing the curious old town of La Roche, two hours' walk from Thorens, the heat and dust of the dreary high road became insupportable; and no pedestrian who undertakes that march with a heavy knapsack, under a blazing noonday sun, will arrive at Bonneville without infinite thankfulness that he has got through it. The road is of the same character as that between Bonneville and Geneva, and that will sufficiently express its unpleasantness in baking times of drought.

The Glacière of the Brezon lies at no great distance from Bonneville--perhaps not more than four or five miles to the SE.--but its elevation is more than 4,000 feet, and the approach is steep. The Glacière of the Valley of Reposoir, a valley which falls into the main road between Bonneville and Chamouni at the village of Scionzier, is considerably higher, and a good deal of climbing is necessary in visiting it. When I arrived at

Bonneville, the whole mass of mountains in which these caves lie was enveloped in thick dark clouds, and the faint roar of thunder reached our ears now and then, so that it seemed useless to attempt to penetrate into the high valleys. Moreover, I was due for an attempt upon Mont Blanc in the beginning of the next week, and an incipient bilious fever, with a painful lameness of one leg, warned me that my powers were coming to an end, and that another day such as the last had been would put a total stop upon the proposed ascent; and so I determined to take the fever and the leg to Geneva, and submit them to medical skill. This determination was strengthened by the exhortations of a Belgian, who called himself a *grand amateurdes montagnes*, on the strength of an ascent of the Môle and the Voiron, and in this character administered Alpine advice of that delightful description which one meets with in the coffee-rooms at Chamouni. This Belgian was the only other guest of the Hôtel des Balances; and his amiability was proof even against the inroads of some nameless species of *vin mousseux*, recommended to me by the waiter, which supplied *mal-à-propos* wine-sauce to the various dishes from which the Belgian was making his dinner, and did not leave his face and waistcoat free from stain. He had but one remark to make, however wild might be the assertions advanced from the English side of the table, '*Vous avez raison, monsieur, vous avez parfait-e-ment raison*!' It is not quite satisfactory to hold the same sentiments, in every small particular, with a man who clips his hair down to a quarter of an inch, and eats haricots with his fingers; but it was impossible to find any subject on which he could be roused to dissentience. This phenomenon was explained afterwards, when he informed me that he was a flannel-merchant travelling with samples, and pointed out what was only too true, namely, that the English monsieur's coat was no longer fit to be called a coat.

Professor Pictet read a paper on these glacières before the *Société Helvétique des Sciences Naturelles* at Berne, in 1822, which is to be found in the *Bibl. Universelle de Genève.*[77] M. Pictet left Geneva in the middle of July to visit the caves, but found himself so much knocked up by the first

[77] Première Série, t. xx. pp. 261, &c.

day's work, that he sent on his grandson to the Glacière of the Brezon, and gave up the attempt himself. The young man found it to be of small dimensions, 30 feet by 25, with a height of 10 or 12 feet. The ice on the floor was believed by the guide to be formed in summer only, and was placed too irregularly to admit of measurement. Calcareous blocks almost choked the entrance, and an orifice in the shape of a funnel admitted the snow freely from above, and was partly filled with snow in July. Cold currents of air proceeded from the rocks in the neighbourhood of the glacière, giving in one instance a temperature of 38°·75, the temperature in the shade being 51°. Within the cave, the temperature was 41°.

M. Morin visited this glacière in August 1828. He describes it as a sheltered hole, in which the snow collects and is preserved.

M. Thury examined it in August 1859, and gives the same account. He, too, found the current of air which the younger Pictet discovered, but in the cave itself the air was perfectly still.

It was clearly, then, no great loss to miss the Glacière of the Brezon; but that on the Mont Vergy, in the Valley of Reposoir, appears to be much more interesting. Professor Pictet found himself sufficiently strong after a day's rest to pass on to Scionzier, and up the Valley of Reposoir, accompanied by the well-known guide Timothée, whose botanical knowledge of the district is said to be perfect. He had conducted MM. Necker and Colladon to the glacière in 1807, and believed that no *savant* had since seen it. The rocks are all calcareous, with large blocks of erratic granite. The glacière lies about 40 minutes from the Châlet of Montarquis, whence its local name of *La grand' Cave de Montarquis*. Before reaching it, a spacious grotto presents itself, once the abode of coiners: this grotto is cold, but affords no ice, and near it M. Morin found a narrow fissure, leading into a circular vaulted chamber 15 feet in diameter, in which stood a solitary stalagmite of ice 15 feet high.

The entrance to the glacière itself is elliptical in shape, 43 feet broad at the base, and the cave increases in size as it extends farther into the rock, the floor descending gently till a horizontal esplanade of ice is reached. This esplanade was 66 feet by 30 at the time of Pictet's visit, deeper in the middle than at the sides, and mounting the rock at the farther side of the cave; there was a small stalagmite at one side, but that would seem to have been the only ornamentation displayed. The temperature was 34°·7, a foot above the ice, and 58° in the external air. Timothée had been in the glacière in the previous April, and had found no ice,--nothing but a pool of water of considerable depth. M. Thury, in August 1859, found two sheets of ice in the lowest part of the cave: one, nearly 50 feet long, was partially covered with water; the other, presenting an area of about 14 square yards, showed more water still. There were no stalactites and columns such as M. Morin had found in August 1828, nor even the low stalagmite which Pictet saw in 1822. The summers of 1828 and 1859 were exceptionally hot, and this fact has been held to account for the smaller quantity of ice seen in those years. M. Thury found the cold due to evaporation to be considerably less than 1° F.,[78] and he and M. Morin both fixed the general temperature of the cave at 36°·5; they also found a current of air entering by a fissure in the lowest part of the cave, but it did not disturb the whole of the interior, for in one part the air was in perfect equilibrium. M. Gampert,[79] in the summer of 1823, found a strong and very cold current of air descending by this fissure, along with water which ran from it over the ice; he believed that this was refrigerated by evaporation, in passing through the thickness of the moist rock.

Two peasants visited this cave three times in the winter season, viz. on October 22, November 26, and on Christmas Day; and one of them, by name Chavan, drew up an account of their experiences, which was read by M. Colladon before the *Société de Physique et d'Histoire Nat. de Genève* in 1824.[80] The peasants found very little ice in columns at the time of the October visit, and there were signs of commencing thaw. The thaw was much more pronounced in November, when the ice had nearly disappeared even from the lowest parts of the cave, and they found the air within quite warm. On Christmas Day they had great difficulty in reaching the glacière, and narrowly escaped destruction by an avalanche, which for a time deterred them from prosecuting the adventure: they persisted, however,

---

[78] Less than 1/2° C., he says.

[79] *Bibl. Univ. de Genève*, Première Série, t. xxv. pp. 224, &c.

[80] *Bibl. Univ.* l.c.

and were rewarded by finding only water where in summer all was ice, and a temperate warmth in the cave. They observed that the roof had fissures like chimneys.

This account was so circumstantial, that the only thing left was to attempt an explanation of the phenomena reported, and such explanations have not been wanting. But M. Thury was not quite satisfied, and he determined to visit the cave in the winter of 1860-1. Accordingly, accompanied by M. André Gindroz, who had already joined him in his unsuccessful attempt to reach the Glacière of the Pré de S. Livres, he left Geneva on the 10th of January, and slept at the Chartreuse in the Valley of Reposoir. As the party passed through the village of Pralong du Reposoir, the peasants told them with one accord that they would find nothing but warmth and water in the cave; but when M. Thury asked had any of them seen it themselves, they were equally unanimous in saying no, explaining that it was not worth anyone's while to go in the winter, as there was no ice to be seen then,--a circular line of argument which did not commend itself to the strangers.

At the very entrance of the grotto, they found beautiful stalactites of clear ice; and here they paused, till such time as they should be cool enough to enter, for the thermometer stood at 70° in the sun, and their climb had made them hot. On penetrating to the farther recesses of the cave, where the true glacière lies, they found an abundance of stalactites, stalagmites, and columns of ice, with flooring and slopes of the same material: not a drop of water anywhere. The stalagmites were very numerous, but none of them more than three feet high; some of the stalactites, fifteen or so in number, were six or seven feet long, and there were many others of a smaller size. M. Thury was particularly struck by the milky appearance of much of the ice, one column in particular resembling porcelain more than any other substance. This is a not unusual character of the most beautiful part of the decorations of the more sheltered ice-caves, as for instance the lowest cave in the Upper Glacière of the Pré de S. Livres; the white appearance is not due to the presence of air, for the ice is transparent and homogeneous, and the naked eye is unable to detect bubbles or internal fissures.

The temperatures at 1.25 P.M. and 2.12 P.M. respectively were as follows:--In the sun, between 3 and 4 feet above the snow, 72°·1 and 70°·5; in the shade, outside the cave, 36°·7 and 35°·8; at the Observatory of Geneva, in the shade, 27°·3 and 28°·2, having risen from 24°·5 since noon. In the cave, 1 foot above the surface of the ice-floor, the thermometer stood at 24°·8; and in a hole in the ice, some few inches below the surface, 24·1. In the large fissure, which has been already mentioned as the source of the summer currents of air, the temperature at various points was from 29°·3 to 27·5. The circumstances of these currents of air were now of course changed. Instead of a steady current passing from the fissure into the cave, and so out by the main entrance into the open air, strong enough to incline the flame of a candle 45°, M. Thury found a gentle current passing from the cave into the fissure, sufficient only to incline the flame 10°, and near the entrance 8°, while in the entrance itself no current was perceptible at 4 P.M.

M. Thury remarks that less current was to be expected in winter than in summer, because the upper ends of the fissures would be probably choked with snow, and their lower ends with ice. It is evident that the current which passes up into the fissure in winter, is favourable to the introduction of the colder air from without; while the opposite current in summer keeps up a supply of cold air in the cave, and so increases its powers of resisting the attempts of the heated external air to make a partial entrance. Both these currents, then, favour the glacial conditions of the cave, and to some extent counterbalance the disadvantages of its situation: viz., its aspect, towards the south-east; the large size of its opening to the air, and the absence of all shelter near the mouth, such as is so often provided by trees or rocks. The small depth of the cave, scarcely amounting to 18 feet below the level of the entrance, is also a great disadvantage.

The people of Pralong asked, on the return of the party, what had been found in the *grand' cave*, and the answer reduced them to silence for a few moments. Their prejudices, however, were invincible, and they persisted in their belief that a true glacière ought to have no ice in it in the winter. M. Thury did not enquire from what source they drew their ideas of a true glacière.

There is a book, in three volumes, on the 'Glacières of the Alps,' by M. Bourrit, dedicated to Buffon, in which is a description of the Valley of Reposoir; but no mention whatever is made of the *grand' cave*. Indeed, M. Bourrit merely meant by *glacière*, a glacial district, something more extensive than a *glacier*, and he had evidently no knowledge of the existence of caves containing ice.

Chapter 13

La Borna De La Glace, in the Duchy of Aosta

The Chanoine Carrel, of Aosta, whose name is so well and so favourably known to Alpine men, sent a brief account of an ice-cave in his neighbourhood to the *Bibliothèque Universelle* of Geneva[81] in the year 1841, and, as far as I know, there is no other account of it. My plan had been to pass from Chamouni by the Col du Géant to Courmayeur, and thence to Aosta for a visit to the canon and his glacière; but, unfortunately, the symptoms which had put an end to the expedition to the Brezon and the Valley of Reposoir came on with renewed vigour, as a consequence of Mont Blanc, and the projected fortnight with Peter Pernn collapsed into a hasty flight to Geneva. It was fortunate that medical assistance was not necessary in Chamouni itself; for one of the members of our large party there was mulcted in the sum of £16, with a hint that something beyond that would be acceptable, for an extremely moderate amount of attendance by the local French doctor.

The glacière was thus of necessity given up. It is known among the people as *La Borna de la Glace*, and lies about 5,300 feet above the sea, on the northern slope of the hills which command the hamlet of Chabaudey,

---

[81] Nouvelle Série, t. xxxiv. p. 196.

commune of La Salle, in the duchy of Aosta, to the north-east of Larsey-de-là, in a place covered with firs and larches, and called Plan-agex. The entrance has an east exposure, and is very small, being a triangle with a base of 2 feet and an altitude of 2-1/2 feet. After descending a yard or two, this becomes larger, and divides into two main branches, with three other fissures penetrating into the heart of the mountain, too narrow to admit of a passage. The roof is very irregular, and the stones on the floor are interspersed with ice, which appears also in the form of icicles upon the walls; and, in the eastern branch of the cave, there is a cylindrical pillar more than 3 feet long, with a diameter of rather more than a foot. The temperature at 4 P.M. on July 15, 1841, was as follows:--The external air, 59°; the cave, at the entrance, 37·2°; near the large cylinder, 35°·7; and in different parts of the western branch, from 33°·6 to 32°·9.

M. Carrel was evidently not aware of the existence of similar caves elsewhere. He recommends, in his communication to the *Bibliothèque Universelle*, that some scientific man should investigate the phenomena, and explain the great cold, and the fact of the formation of ice, which common report ascribed to the time of the Dog-days. He doubts whether rapid evaporation can be the only cause, and suggests that possibly there may be something in the interior of the mountain to account for this departure from the laws generally recognised in geology.

Chapter 14

THE GLACIÈRE OF FONDEURLE, IN DAUPHINÉ

There cannot be any better place for recruiting strength than the lovely primitive valley of *Les Plans*, two hours up the course of the Avençon from hot and dusty Bex. Here I rejoined my sisters, intending to spend a month with them before returning to England; and the neighbouring glaciers afforded good opportunities for quietly investigating the structure of the ice which composes them, with a view to discovering, if possible, some trace of the prismatic formation so universal in the glacières. On one occasion, after carefully cutting steps and examining the faces of cleavage for an hour and a half, I detected a small patch of ice, under the overhanging rim of a crevasse, marked distinctly with the familiar network of lines on the surface; but I was unable to discover anything betokening a prismatic condition of the interior. This was the only case in which I saw the slightest approach to the phenomena presented in ice-caves.

There remained one glacière on M. Thury's list, which I had so far not thought of visiting. It was described as lying three leagues to the north of Die in Dauphiné, department of the Drôme, at an altitude of more than 5,000 feet above the sea. M. Héricart de Thury discovered this cavern in

1805, and published an account of it in the *Annales des Mines*[82] to which M. Thury's list gave a reference. I have since found that this account has been translated into various scientific periodicals, among others the Philosophical Journal of Edinburgh.[83] earlier than I had intended, I could take advantage of the new line connecting Chambéry and Grenoble and Valence, and so visit this glacière without making the journey too long; and accordingly I bade farewell to Madame Chérix's comfortable room, leaving my sisters in their quarters in a neighbouring châlet, and started for Geneva.

The line was advertised to open on the 15th of August; but on the 16th the officials declared that it was not within a month and a half of completion, so that I was compelled to go round by Lyons. I was easily reconciled to this by the opportunity thus afforded of a visit to the ancient city of Vienne, which well repays inspection. Its history is a perfect quarry of renowned names, Roman, Burgundian, and ecclesiastical. Tiberius Gracchus left his mark upon the city, by bridling the Rhône--*impatiens pontis*--with the earliest bridge in Gaul: and here tradition has it that the great Pompey loved magnificently one of his many loves; while the site of the Prætorium in which Pontius Pilate is said to have given judgment can still be pointed out. The true Mount Pilate lies between Vienne and Lyons, being one of the loftiest northern summits of the Cevennes, on the borders of the Lyonnaise.[84] The Romans recognised the fitness of the neighbourhood of Vienne for the cultivation of the grape, and the first vine in Gaul was planted on the Mont d'Or in the second century of the Christian era. In Burgundian times the city held a very prominent place, and became infamous from the frequent shedding of royal blood; so that early historians describe it as '*tousiours fatale à ceux qui vueillent la corone des Bourgougnons,*'[85] and as '*fatale et de malencõtre aux tyrãs et mauvais princes.*'[86] Ecclesiastically, its interest dates of course from a very early period, from the times of the martyrs of Gaul and the first Rogations.

[82] T. xxx. p. 157.
[83] Vol. ii. p. 80.
[84] Jean de Choul, *De variâ Quercûs Historia*, 1555.
[85] Gollut, Mém. des Bourg. de la Franche Comté, p. 227.
[86] Paradin de Cuyseaulx, Annales de Bourgougne, 1566, p. 14.

The Festival of *Les Merveilles* long commemorated the restoration of the bodily forms of the Lyonnese martyrs, as their scattered dust floated past the home of Blandina and Ponticus; and the dedication of the cathedral to S. Maurice keeps alive the tradition that Paschasius, bishop of Vienne, was warned by an angel to watch on the banks of the Rhône, and so rescued the head and trunk of the soldier-martyr, which had been cast into the river at Agaunum (S. Maurice in Valais), and had floated down--probably on sounder hydrostatical principles than the 'Floating Martyr'--through the Lake of Geneva, and so to Vienne. There are still many very interesting Roman remains in the city, as the Temple of Augusta and Livia, the Arcade of the Forum, and the monument seen from the railway to the south of the town. The temple is being carefully restored, and the large collection of Roman curiosities which it contained is to be removed to the church of S. Peter, now in course of restoration, which will in itself be worth a visit to Vienne when the restoration is completed.[87] All the buildings connected with the Great Council in 1311 have disappeared; and the only relic of the council seems to be the Chalice, *or*, surmounted by the Sacred Host, *argent*, in the city arms, in remembrance of the institution of the Fête of the *S. Corps*. If the Emperor would but have the town and its inhabitants deodorised, few places would be better worth visiting than Vienne.

The poste leaves Valence--the home of the White Hermitage--for Die at 2.30 P.M., and professes to reach its destination in six hours; but sad experience showed that it could be unfaithful to the extent of an hour and a half. So long as the daylight lasted, there was no dearth of objects of interest; but when darkness came on, the monotonous roll of the heavy diligence became aggravating in the extreme. The village of Beaumont, once the residence of an important branch of the great Beaumont family,[88] retains still its square tower and old gateway; and the remains of a château near Montmeyran, the end of the first stage, mark the scene of the victory

---

[87] Several churches in Vienne are used as foundries and workshops. S. Peter's church was an iron-foundry four or five years ago, and is in future to be a museum—a considerable improvement upon its former use. The grand old church of S. John in Dijon has been rescued from the hands which made it a depôt of flour, and is being restored to its original purposes: but such instances are very rare.

[88] This family took its rise in Dauphiné, before the district had that name: the chief place of the family was the château of Beaumont, near Grenoble.

of Marius over the Ambrons and Teutons, local antiquaries believing that the name of Montmeyran is from *Mons Jovis Mariani*.[89] The road lies through the bright cool green of wide plantations of the silkworm mulberry,[90] with its trim stem and rounded head; and, in the more open parts of the valley, walnut trees of size and shape fit for an ornamental park in England relieve the monotony. The nearer hills are covered to the top with vines, and the higher and more distant ranges have a naked and thoroughly burned appearance, which suggests the idea of volcanoes to a traveller ignorant of volcanic facts. The villages which lie at the foot of these rocky hills are built of stones taken from the beds of the streams, and are so completely of one colour with the background of rock, that in many instances it is difficult to determine whether a distant mass of grey is a village or not. Ruined castles and towers abound; and these, and still more the walls which surround many of the villages, point unmistakeably to times of great disturbance. The valley of the Drôme, up which the road after a time turns, was an important locality in the religious wars; and the town and fort of Crest especially, as its name might suggest, was a famous stronghold, and resisted all the efforts of the Reformed party. In yet earlier times, Simon de Montfort had frequently tried to take it, without success; and four years after S. Bartholomew, Lesdiguières met with a like repulse.[91] The same80 story of sieges and battles might be told of almost every village and defile of the valley. Thus, Saillans, the third stage, was taken by the Protestant leader Mirabel, and the Catholic Gordes, in 1574, and its fortifications were razed by the Duc de Mayenne in 1581. Pontaix, again, a remarkable place, with a vaulted street and fortified houses

---

[89] The final victory was near Aquæ Sextiæ (Aix).

[90] The cultivation of the silkworm mulberry will probably die out before very long. The silk crop has lately failed in Dauphiné, and a commission for enquiring into the relative merits of different worms has determined that the Senegal worm produces 633 millegrammes of silk, while the worm, fed on the mulberry produces only 290. The first mulberry trees in France were planted in that part of Provence which is enclosed by Dauphiné.
The Bishop of Nismes has lately issued a pastoral letter, commanding prayers to be offered up for the cessation of the malady affecting the silkworms in his own and the surrounding dioceses.

[91] The feudal buildings were razed by order of Richelieu, but the tower remains a landmark for the valley. Three hundred *détenus* were confined here after the *coup d'état* of December 2, 1851.

overhanging the river, which here fills up the whole valley and leaves room only for the road and the narrow village-town, was the scene of an obstinate and murderous fight between the Marquis de Gordes on one side, and Lesdiguières and Dupuy-Montbrun on the other, when the latter was captured, and shortly after beheaded at Grenoble.

The town of Die, *Dea Vocontiorum*, lies in a broad part of the valley. It claims to be not *Dea Vocontiorum* only, but also *Augusta Vocontiorum*, thereby apparently defrauding the village of Aouste, near Crest, of the earliest form of its name. Die is possessed of old walls, and has four gates with towers. The great goddess from whose worship it derives its name was Cybele, notwithstanding the vehement assertions of the official in the Poste-bureau in favour of Ceres; and three different Tauroboles have been discovered here, one of which is in excellent repair, and shows a Roman inscription surmounted by three bulls' heads. The ceremony of the Taurobolium was new to me, and appears to have been conducted as follows:--A small cave was hollowed out, with a thin roof formed by the outer surface of the earth; and immediately above this a bull was sacrificed, so that the blood ran through the earth and dropped on to a priest who was placed in full robes in the cave. The priest and the blood-stained garments were thenceforth specially sacred, the garments retaining their sanctity for twenty years. The inscription on the Tauroboles which have been found in and near Die record the names of the priest, the dendrophore, the person who provided the victim, and the emperor for whose safety the sacrifice was offered.

The people of Die have been quarrelsome from the earliest times. A century before the estates of the Dauphins of the Viennois were known as Dauphiné,[92] the chronic contests between the Bishops and the Counts of Die had come to such a crisis, that the Dauphin Guiges André intervened, and produced a certain amount of peace; but, twenty years after, the people killed Bishop Humbert before the gate which thence received its name of *Porte Rouge*. When the Counts of Valentinois had succeeded to the fiefs of

[92] The origin of the name Dauphin seems to be lost in obscurity, though of comparatively recent date. The Counts d'Albon took the title first in 1140, and their estates were not called the Terra Dalphini, or Dalphinatus, till 1291. The first Dauphins bore a castle, not a dolphin.

the Counts of Die, Gregory X. became so weary of the constant wars, that he suppressed the bishopric, and united it to Valence in 1275; but the canons, who were not suppressed, raised a mercenary army and carried on the struggle. Eventually, the canons and the people made common cause, and joined the Pope during the Seventy Years; but when he left Avignon they came to terms with Charles VI of France, and so the Diois was united to Dauphiné in 1404. Louis XIV restored the separate bishopric, but ruined the town by the revocation of the Edict of Nantes.

The large number of mosaics and inscriptions found in Die prove conclusively that in Roman times it was a favourite place of residence; and, so far as situation goes, it is not difficult to understand how this should have been the case. But in the condition in which the town found itself in the pitiless heat of August 1864, the only question for an English visitor was whether he could live through the time it was absolutely necessary to spend there. The poste arrived, as has been said, an hour and a half after its time; and the sole occupant of the coupé, who had lived on fruit and gooseberry syrup, and three penny worth of sweet cake at Crest, since a seven-o'clock breakfast, had wiled away the last hour by inventing choice bills of fare for the meditated supper. When the lumbering vehicle stopped in the main street of Die, which is here something under seven yards wide, an elderly woman stepped out from the dim crowd, with an uncovered tallow candle in her hand, and asked if there was anyone for the hotel. The unwonted 'yes' seemed to create some surprise; but she led the way promptly to her hotel, diplomatically meeting the rapid volley of questions respecting supper with an unpromising silence, and the first sight of the house itself dispelled forever all hope. An entrance was effected by the kitchen; and not only was there no fire, but there was no light of any description; and the one dip we brought on to the scene betrayed such squalor on all sides, that the suggestion of a *salle-à-manger* in connection with such a kitchen became at once an impudent mockery. When this farther room was reached, it proved to be even worse than the kitchen. It was shut up for the night--had been shut up apparently for a week--and was in the possession of the cats of the town, and the flies of Egypt. Two monstrous hounds entered with us; and the cats fled hastily by a window which was slightly open at the top, spitting and howling with fear when they missed the first spring, and came within the cognisance of their mortal foes.

The first thing to be done was to wash off some of the accumulated dust; but when I asked for a bedroom for that purpose, I was conducted to a copper in the kitchen, the water in which had been a permanency for some time past, and was told to wash there. As for supper, there was some cold mutton; but the landlady unfortunately opened the door of the cupboard as she said so, and displayed a state of things which decided the point against the mutton. There was nothing else in the house, and there was no fire for cooking anything; but when they discovered that I really would not wait till the next morning, they proposed to light a fire and warm some soup, which I declined to see in its present state. In the way of wine, I had been recommended to make a great point of the *clairette de Die*, an excellent species of *vin mousseux*; but the chief of the women rather recommended the ordinary wine of the country, as the monsieur might not like to give a strong price. 'Was it, then, so strong?' 'Yes, the price was undoubtedly strong.' 'How much, then?' 'A franc a bottle.' With an eye to the future bill, the monsieur pretended to ponder awhile, as if in doubt whether his resources could stand such a strain, and then, with a reckless air, decided upon the extravagance. The clairette proved to be quite worthy of the praise which had been bestowed upon it, being a very pleasant and harmless sparkling white wine.[93]

The neighbours kept dropping into the kitchen, to see how the landlady got on with the stranger of uncouth speech; and four of the female part of her company brought in at various times to the *salle-à-manger* some piece of table-furniture, in order to indulge in a closer view than the open door of the room afforded. One of them told me she had seen an Englishman once before, a few months back; but he only had one eye, and she seemed to think I was out of order in possessing two. At length the soup came, and the first attempt upon it proved it to be utterly impossible. The landlady was called in, and this fact was announced to her. 'What to do, then?--it

[93] The old historian Gollut speaks of the *clairets* and *clerets* as red wines.

was a good soup, a soup which the people of Die loved,--it was a soup the household eat morning and night.' All the same, it was not a soup the present Englishman could eat, and some other sort of food must be provided, for she declined to furnish soup without garlic and fat. She suggested an omelette; but a natural generalisation from all I had so far seen drew an untempting picture of the probable state of the frying-pan, and I declined to face the idea until I was convinced there was nothing else to be had. But, alas! notwithstanding the righteous indignation with which the landlady met my request that the omelette might not be all fat, the manipulation of the eggs eventuated in a dish even more impracticable than the soup, flooded with unmentionable grease, and so at last the cold mutton became a necessity. To show how hunger may work upon the feelings, I may say that, in spite of the marks of the feet of mice in the cold gravy which remained on the dish, I forced myself to cut off a wedge, and, after removing a thick layer of meat on the exposed sides, essayed to eat the heart of the wedge. The sheep and its progenitors had been fed on garlic from all time, and the mutton had been boiled in a decoction of that noxious herb; and this dish was in its turn rejected like the others. There was nothing for it but salad, and bread, and wine; but when the salad appeared, after a long time had been spent in the kitchen in saturating the withered greens with oil and vitriolic vinegar, there, perched on the top like one of those animals which sometimes spoil one's enjoyment of a strawberry-bed, was a huge onion, with numerous satellites peeping out from under the leaves. About this time, a short diversion was caused by the reappearance of one of the large hounds, whose mind was not at ease as to the completeness of the previous elimination of the cats from the *salle-à-manger;* and the diabolical noise and scuffle which ensued upon his investigation of a dark corner, showed that his doubts had been well grounded. Then I discovered that there was no butter to be had, and no milk; and when coffee was mentioned, a pan was brought out for making that beverage, which a bullet-maker with any regard for appearances would have declined to use for melting his lead in. Finally, under the pressure of dire hunger, I returned to the mutton, and contrived to swallow a small piece, the taste of which did not leave me for four or five days.

The interior of the house, where the bedrooms were, gave forth an odour which must be familiar to all who have burrowed in out-of-the-way places in France, approaching more nearly, perhaps, to the smell of damp cocks and hens than anything else; and the bedroom door was guarded by a huge mis-shapen dog, which evidently intended to pass the night there, if it could not get into the room itself. The street on to which the window looked was still populous with the inhabitants of Die; and a man with whom I had already had a conversation respecting the glacière, who appeared to perform some of the functions of landlord of the hotel, was audibly engaged in hiring a man to accompany me on the following day. The man whom he was attempting to persuade was evidently of an independent turn of mind, and said that as it would be an affair of fifteen or sixteen hours at least, he would not go through so much unless his proposed comrade were a true *bonhomme*; a difficulty which the landlord set at rest by asseverations so ready and so circumstantial, that I determined to take everything he might tell me, on any subject, with many grains of allowance.

It was only natural to expect a night of horrors; but in this I was most agreeably disappointed, and the few hours passed quietly enough till it was time to get up. By morning light, the *salle-à-manger* did so bristle with squalor that the kitchen was made the breakfast-room; though as that meal only lasted two minutes, and meant nothing beyond an attempt to eat some of the bread I had been unable to eat the night before, one place was much the same as another. It is generally believed that coffee is to be obtained in perfection in France; but that belief is not founded on experience of the provinces, and had long ceased to be a part of my creed: nevertheless, with the idea that there is always some redeeming-point in the darkest situation, I had hopes of the coffee of Die, in spite of the appearance of the pan; and if these hopes had been realised, the place might still have been tolerable. But they were not realised. When the landlady was asked for the promised coffee, she brought out a small earthenware pitcher containing a black liquid, and proceeded to bury its lower extremity in the hot embers of the wood fire, by which means the liquid was speedily warmed up, and also thickened with unnecessary ashes. When served--in the same dusty pitcher-

-it had a green and mouldy taste, combined with a sour bitterness which made it utterly impossible as an article of food, and so the breakfast was confined to the rejected fragments of the loaf of the preceding night.

The guide, or comrade as he preferred to call himself, appeared in good time, and we started about half-past six, under a sun already oppressively hot, and through heavy flaky dust, which made us feel very thankful when our route branched off from the high road. Liotir was strong in mulberry trees and vines, for he was a keeper of silkworms, and a wine-merchant. Silkworms had not been profitable for a year or two, and he was almost in low spirits when he talked of them.[94] An epidemic had visited the district, and the worms ate voraciously and refused to spin--a disease which he believed to be beyond the power of medicine.[95] As is so often the case with the Frenchman, as compared with the Englishman of corresponding social status, he had his information cut and dried, and poured it out without hesitation. Silkworms' eggs cost 15, 20, or 25 francs an ounce, according to quality; and an ounce of good seed should produce from two to three hundred francs' worth of cocoons. A man who 'makes' an ounce of seed requires six tables, 8 feet by 4, for his cages; and as some men make thirty-five ounces, chambers of great size are necessary for the accommodation of their worms; but breeders to so large an extent as this are the princes of the trade. As we passed a farmhouse surrounded by mulberry trees and vineyards, my companion informed me that the farmer was his partner in worms and wine both, and that the wine promised to be the better speculation this year, for the fruit was in immense abundance. I saw afterwards that, at the time of vintage, grapes sold for pressing at from 6 to 10 francs the hundred kilos, while 12 and 13 francs was the price in 1863, and that in some districts of the Drôme the owners of the presses had not barrels enough for even the first pressing.

---

[94] The 'Times' of Oct. 4, 1864, stated that almost no raw silk was offered at the last markets at Valence and Romans, and but for foreign supplies the mills must have been closed. The small amount that was offered sold at from 68 to 72 francs the kilogramme, while foreign cocoons from Calamata fetched only 22 francs at Marseilles.

[95] Pausanias says that silkworms are apt to die of indigestion, the cocoons lying heavy on the stomach.

The great want of wood on the hills in whose neighbourhood we now found ourselves, attracted attention in the time of Louis XIV., and that sovereign passed severe laws for the protection of the forests that still remained. As usual, the mere severity of the laws made them fail of their object. Banishment and the galleys were the punishment for unauthorised cutting of forest trees, and death if fire were used. There is a paper in the *Journal de Physique* of 1789,[96] on the disappearance of the forests of Dauphiné, pointing out that when the woods are removed from the sides of mountains, the soil soon follows, and the district becomes utterly valueless. The writer traced the mischief to the emancipation of serfs, and the consequent formation of *communes*, where each man could do that which was right in his own eyes.

At any rate, whatever the reason, nothing can be conceived more bare than the dun-coloured rounded hills between the town of Die and the Col de Vassieux, towards which we were making our way. The whole face of the country had the same parched look, and the soil seemed to be composed entirely of small stones, without any signs of moisture even in the watercourses. The Col de Vassieux is not much more than 4,000 feet high, and forms a saddle between the Pic de S. Genix (5,450 feet) and the But de l'Aiglette (5,200 feet). A new foot-road has been made to the Col, with many windings; and great care has been taken to plant the sides of the hill with oak and hazel; so that already there is some appearance of coppice, and in the course of time there will be shade by the way--a luxury for which we longed in vain. The lower ground was covered with little scrubs of box, and with lavender, dwarfed and dry; but near the summit of the Col the lavender became vigorous and luxuriant, and carpeted the hillside with a rich abundance of blue, tempting us more than once to lie down and roll on the fragrant bed; though some of the older roots were not sufficiently yielding to make that performance as satisfactory as it might have been. This lavender is highly prized by the silkworm-keepers of Die, its bushy heads being almost exclusively used for the worms to spin their cocoons in.

---

[96] T. xxxv. pp. 244, &c.

When we reached the top of the Col, Liotir confessed that he did not know which way to turn, and we agreed to follow the path till we should find some one to direct us. There was a farmhouse at no great distance, and thither we bent our steps; but the sole inhabitant could give no assistance, and, in default of information, Liotir generously proposed to treat me to a bottle of wine, over which we might discuss our further proceedings. The state of fever, however, to which the garlic and the dirt of Die had brought me, made it seem impossible to eat or drink anything; so I suggested instead that I should treat him, and that seemed to be rather what he had meant by his proposal. Nothing much came of our discussion, and we marched on hot and faint for an hour more, when a casual man told us that our straight line to the *Foire de Fondeurle* lay across the plain on our left hand, and up a most objectionable-looking hill beyond, thickly covered with brushwood and showing no signs of a path.

As we crossed the plain, there was still the same total absence of water, and we reached the bottom of the hill in a state of mind and body which rebelled against the exertion of struggling with the sand and shingle and brushwood. Liotir thought it was useless to attempt it with no hope of water, and I held much the same view, only it was impossible really to think of giving it up. When at last we had surmounted all the difficulties which beset us, and stood on the highest point which had so far been in sight, we found ourselves on the edge of a vast plain of parched grass, with nothing to guide us in one direction rather than another. There was no human being in sight, no sign of water, nor any particle of shade; nothing but grass, brown and monotonous, with white cliffs miles away at the extremity of the plain. This was evidently the *Foire de Fondeurle*, and in it somewhere lay the glacière, if only we could make out in which direction to begin to traverse the plain. In the earlier part of this century, a very famous fair was held on this wild and out-of-the-way table-land, to which many thousands of horses and mules and cattle of various kinds were brought from all quarters; but the fair has fallen off so much, that the man who had turned us up the last hill said there were only fourteen head of cattle in 1863, and very few of those were sold. M. Héricart de Thury describes this plain as lying in the calcareous sub-Alpine range of the south-east of France. The woods here terminate at a height of 5,147 feet above the sea, and the *Foire de Fondeurle* lies immediately above this point.

At last we made a bold dash across the plain, and after a time came upon some sheep, standing in a thick row, with their heads thrust under a low bank which afforded a little shade; and at no great distance from them sat the shepherd. He was a cripple, and his clothes were something worse than rags. He offered us a portion of the water he had in a detestable-looking skin; but he assured us it was quite warm, and had not been good to begin with, so we did not try it, though we were thirsty enough to have hailed a muddy pool with delight. Our new acquaintance knew nothing of the glacière, but he belonged himself to the Chalêt of Fondeurle, and as that was the only house on the whole plain, he told us to make for it. The surface of the plain seemed to have fallen through in many places, forming larger and smaller pits with steep sides of limestone. These were often of the size of a large field, and, as the deeper of them required circumvention, the shepherd told us that we must follow the line of little cairns which we should find here and there on our way, the only guide across the plain. He could not be sure himself in what direction the châlet lay; but if we kept to a certain tortuous line, we should come to it in time.

The way proved to be so very long, that we doubted whether such a consummation of our wishes would ever arrive: but at length, in a small dip at the farthest extremity of the plain, we saw the châlet, and, what was much more to us, saw a little run of water, carried from the rising ground by wooden pipes. It will be well for any future visitor to the châlet to go very warily, and to intrench himself in a strong position when he sees half-a-dozen huge dogs like black and white bears come out to attack him. Liotir had a stout stick, and I had a formidable ice-axe; and, moreover, we fortunately secured a wall in our rear: but with all this the dogs were nearly too much for us, and Liotir was pressing me earnestly to chop at the ringleader's head, when a man came and called off 'Dragon,' and the others then dispersed. The new-comer wished to know our business, but, without satisfying his curiosity, we rushed to the water-trough, and drank and used in washing an amount of water which he evidently grudged us. Then we were able to tell him that our business was something to eat for Liotir, and a guide to the glacière; though I trembled when I suggested the latter, for, after all our labours, I had a sort of fear that the cave would prove a myth. On this point the man cleared away all doubts at once,--we could certainly have a guide, as the *patron* would be sure to let one of them go with us. As to food, there was more doubt, for the master was not yet at home, and his wife would not be able to give us an answer without consulting him. The wife confirmed this statement: they saw very few strangers, and did not profess to supply food to people crossing the plain. I assured her that we intended to pay well for anything she could let us have, but she merely rejoined that they did not keep an auberge; however, her husband would be home sometime in the course of the afternoon--it was now about half-past twelve--and she could ask his opinion on the subject. But Liotir objected that he was meanwhile dying of hunger, and the monsieur of thirst which only milk or cream could assuage; he suggested that someone should be sent to look for the husband, and obtain his permission for us to be fed. To this she assented, very dubiously, and with a constrained air, as if there were some mysterious reason why the presence of strangers was peculiarly unacceptable on that particular afternoon. At any rate, she said when pressed, she thought there could be no harm in our entering the châlet and sitting down on a bench, where we should be sheltered from the sun.

Here accordingly we sat, more or less patiently, till the master himself appeared. He had no welcome for us; but he was willing that we should eat some of his black bread, and try his wine. Liotir begged for cheese, and the wife was told she might supply cheese of two kinds, and also cream, for the monsieur evidently was *malade* and could not swallow wine. The cream and the black bread were delicious; but still the horrors of Die hung about me, and I could only dispose of such a small amount, that Liotir waxed funny, and told me it would never do for me to die there, as there was not earth enough to scrape a grave in on the whole plain. Then, being a practical man, he declared he should like to contract for my keep, and thought he could afford to do it at very small cost to me, and still leave a fair margin for himself. He thought it right to make up for my want of appetite; and so, in addition to his own share, he took in an exemplary manner the share of wine which I should have taken, had I been a man like himself. The master of the châlet sat on the family bed, smoking silently and sullenly; and as soon as Liotir had come to an end of his second bottle, he proposed to accompany us himself to the cave, as he doubted whether any of his men knew the way, and he was sure they were all busy. When I came to pay his wife for what we had consumed, I administered thanks as well as money; to which she sternly rejoined, 'Who pays need not give thanks;' and to that surly view she held, in spite of my attempts to soften her down. There was, after all, much force in what she said, under the circumstances. They had given us no welcome, nothing but mere food, and all they expected in return was a due amount of money; thanks were a mockery in their eyes.

The cavern was reached in a few minutes, when once we got away from the châlet. Two large pits, formed apparently by the subsidence of the surface, lay in a line about east and west, and there proved to be an underground communication between them. From this tunnel, as it were, a long low archway led to a broad slope of chaotic blocks of stone, down which we scrambled by the aid of such light as our candles afforded. The roof of this inner cave was horizontal for some distance, and then suddenly descended in a grand wall; and in consequence of a series of such inverted steps, the cave never assumed any great height. The whole length of the slope was 190 feet, and its greatest breadth about 140 feet; but the breadth varied very much. Half-way down the slope the ice commenced, fitfully at first, and afterwards in a tolerably continuous sheet. The most careless explorer could not have failed to notice the polygonal figures stamped upon its surface. They were larger and bolder than any I had seen before; and the prismatic nuts into which the ice broke, when cut with the axe, were of course in proportion larger than in the previous caves. The signs of thaw, too, were unmistakeable. Though the upper surface of the earth had seemed to be utterly devoid of moisture of any kind, large drops fell freely from the roof of the cave,[97] and the ice itself was wet. The *patron* said

[97] M. de Thury calculated that the thickness of the roof at the lower part of the cave was about 60 feet of rock. He also noticed the peculiar structure of the ice, which afforded great surprise to his party. It was discovered by means of the coloured rays which were thrown into the

there was no ice whatever in the winter months, and that from June to September was the time at which alone it could be found. He declined to explain how it was that we found it so evidently in a state of general thaw in the very height of its season. To give us some idea of the climate of the plain in winter, he informed us that the snow lay for long up to the top of the door of his châlet.

There were in all four columns of ice in the cave, only two of which were of any considerable size. One of these was peculiarly striking from the very large grain which its structure displayed; it measured 19 feet across the base, being flat towards the extremity of the cave, and round towards the entrance. Three thermometers in various parts of the glacière gave all the same temperature, namely, a fraction under 33° F.: a rough French thermometer gave 1/2° C. The extreme wall of the cavern was completely covered by a layer of stalagmitic material, and some of the forms the substance assumed were sufficiently striking. In contact with the wall, though standing clear of it in parts where the wall fell inwards, stood a thick round column of the same material, shaped like the ordinary ice-columns of the glacières, with a cavity near the base, and in all ways following the usual laws of such columns. Considering that I had observed a layer of limestone-paste collecting on one of the ice-columns of the Glacière of La Genollière, I could not help imagining that this stalagmitic column had been originally moulded on a norm of that description. It had a girth of 12 feet in the part where we were able to pass the tape round it. Its surface was smooth; but when we drove a hole through this, with much damage to the *pic* of my axe, we found that the interior was in a crystalline form.

There was, on the whole, very little to be seen in the glacière. Had it been my first experience of an ice-cave, it would doubtless have seemed very remarkable, as it did to Liotir, who, by the way, had steadily disbelieved the possibility of natural ice in summer except in the glaciers; but as I had now seen so many, several of them much more wonderful than this, I did not care to stay longer than was absolutely necessary for

---

different parts of the cave, when someone had casually placed a torch in a cavity in one of the columns.

measurements and investigation. Besides, the food of Dauphiné rather takes the energy and love of adventure out of an unaccustomed visitor.

Without long delay, then, we bade farewell to the *patron*, not returning to the inhospitable châlet, and started on our way for Die, each carrying a large block of ice slung in a network of string. Liotir's purpose was to convince some mysterious female friend that he really had seen ice in summer, within five or six hours of Die; and mine, to apply the ice to the butter which I had specially ordered the landlady to have ready for me, that so I might be able to get through the night, and leave Die by the diligence the first thing next morning. It was remarkable how well the ice bore the great heat. For long the bulk of the masses we carried seemed scarcely to diminish; and if it had not been for a course of heavy falls as we descended through the brushwood, we should have succeeded in getting a large proportion of it safely to Die. The precision of the prismatic structure also showed itself in a very marked manner; and when we came to a crisis of thirst, which happened at shorter and shorter intervals as the afternoon wore on, we separated the prisms with our fingers from the edges of the ice without any difficulty, and made ourselves more hot and thirsty by eating them.

When we arrived at the farmhouse at the Col de Vassieux, we reaped full benefit from our ice. The wine, which had been hot and heavy and unpalatable in the morning, when we had tried it unmixed, became delightfully refreshing when disguised with an abundance of water and sugar and ice; and Liotir found that contracting for my keep at a low rate would not, after all, secure him the comfortable income he had before calculated. After this refreshment, he became communicative, and told me he had served seven years in the French army, three of which were spent in working on railways. He had fought the Italian campaign, and was full of details of the battle of Solferino, on which occasion his *bataillon* was led on by the Emperor in person. According to his account, four *bataillons* were drawn up for the assault of a tower, and when the first advanced it was swept away to a man. The second met with a like fate, and Liotir was in the third. His officers had all been killed, and a corporal was in command. The Emperor rode up and called to them to advance as far as he advanced. This was about a hundred yards; and then, after halting them for a moment, the Emperor cried, '*Allez, mes enfants! nous ne sommes pas tous perdus!*' sending the fourth *bataillon* close upon their heels. In answer to my question, Liotir said, slowly and solemnly, that he did not think the Emperor was under fire; a few dropping shots reached them while he was yet addressing them, but he believed the Emperor Napoleon was not in the fire at Solferino. I took the opportunity of asking whether he was green on that occasion, as Mr. Kinglake believes that he is in times of personal danger; but my companion utterly scouted the idea, and declared that he saw no man through all that day so cool and capable as the Emperor. Pale he undoubtedly was, but that was his habit. Like all other French soldiers with whom I have had much conversation, Liotir complained of the army arrangements in the matter of food; on all other points he was most amiable, but when he spoke of the extortions of the *cantinière* he completely lost his temper. At a *café*, the soldiers could get their cup for 15 centimes, or 20 with liqueur; whereas the *cantinière* charged a franc, and gave them very bad coffee. Wine, too, which would cost them 60 centimes the kilo in the town, was valued at 2 francs by their grasping enemy. He had an idea that English soldiers are allowed to take their whole pay in money, and spend it as they will; whereas the French foot-soldier, according to his account, gets 25 centimes a day in money, and has everything found except coffee. A young trooper at Besançon was very eloquent on this subject. He represented himself as a man of small appetite and a gay spirit; he could well live on very little solid food, and yet he had as much deducted from his pay on that account as anyone in the army--as much, for instance, he groaned, as a certain stout old warrior who was then reposing on a corn-bin. If he could have drawn all his pay in money, and lived on almost nothing for food, he would have had abundance of sous for cards and tobacco; and what a career would that be!

The blocks of ice were by this time becoming rather small; and as we had now once more reached the region of lavender, we cut a large quantity and wrapped the ice in it, and thus protected it from further thaw. For some time before arriving at the farm where my companion's partner lived, he indulged in praises of the wine which their vineyard produced, and assurances of the safety with which it would perform a journey to England. He urged its excellent *bouquet*, and gave me a card of prices which certainly seemed marvellously reasonable. Finally, he proposed to join me at a bottle of white *muscat*, from the farmer's *cave*, in order that I might have an opportunity of seeing how true was his account of the wine. We seated ourselves accordingly in the farmyard, and drank a bottle of delightful wine at 65 centimes the bottle, clear and sparkling, and with a strong muscat flavour. Liotir combined with it intoxication of a different kind, and showed unmistakeable signs of his determination to take another member of the farmer's household into partnership,--the mysterious friend, in fact, for whose astonishment the ice was intended. The white muscat, they told me, would not keep over the year; but they had a wine at the same price which they highly recommended, and warranted to keep for a considerable number of years. Liotir was very anxious that we should have a bottle of this, for he was confident that I should give them an order if I once tasted it; but we had been in at the death of so many bottles that day, that I declined to try the *muscat rosat*. I have since had a hundred *litres* sent over by Liotir, and find it very satisfactory. It has a rich, clear, port-wine colour, sparkling, and with the true *frontignac* flavour.

The effect of the wine on Liotir was peculiar. In the earlier part of the walk, he had never seen Algeria; but after half a bottle of muscat, he had spent six months in that country, and he enlivened the remainder of the way with many details of his experiences there. We reached Die about half-past seven, and the arrival of real ice was hailed as a marvel. Although I had been sent off so unhesitatingly by the landlord in the morning, it seemed that they none of them knew what a glacière meant. They had determined that we should never reach the *Foire de Fondeurle*, and that if we did, we should find nothing there to repay our toil. As I sat at an open window afterwards, Liotir's voice was to be heard holding forth in a neighbouring café upon the wonders of the day; and among the crowd which is a normal condition of the evening streets of Die, the words *Fondeurle*, *Vassieux*, *Anglais*, *glace*, &c., showed what the general subject of conversation was.

The landlady had obeyed orders, and was provided with butter and bread. The tea was served in an open earthenware pitcher, with the spout at right angles with the handle. There was no cup; but the woman remarked that if monsieur was particular about that, he could turn out the sugar and use the basin, which he did. The milk had a basin to itself; but it had offered so large and tempting a surface to the flies of the town, that it remained untouched. The knife and spoon were imbued with ineradicable garlic, and my own trusty clasp-knife was the only weapon I could use for all table purposes. If it had not been for the ice and the lavender, I think I should never have got away from Die. The former made it possible to eat some bread-and-butter; and of the latter I made a sort of respirator for nose and mouth, which modified the odour of cocks and hens prevailing in the house.

Next morning the diligence was to start early, and, in preparation for the six hours' drive, I ordered two eggs to be boiled for breakfast. As the first proved to have been boiled in tepid water, I requested the landlady to boil the second afresh, which she did in a manner that may partly account for the observed fact that the very eggs of some towns taste of garlic. There was household soup simmering on the fire, reeking with onion and garlic, and many other abominations; and, as if it was quite the right and usual thing to do, she slipped the unfortunate egg into this, and left it there to be cooked. After all, garlic must be cheap as an article of food, for the whole bill amounted only to 7-1/2 francs.

This was the last glacière on my list. It was quite as well that such was the case; for the trials of Dauphiné had been too great, and I should scarcely have been inclined to face further adventures of a like kind.

Chapter 15

OTHER ICE CAVES

THE CAVE OF SZELICZE, OR SZILITZE, IN HUNGARY[98]

Matthew Bell, the historian of Hungary, sent an account of this cavern to England, in the middle of the last century, which was printed in the original Latin in the 'Philosophical Transactions' of 1739-40 (pp. 41, &c.).

This account states that the cave is in the county of Thorn,[99] among the lowest spurs of the Carpathians. The entrance, which faces the north, and is exposed to the cold winds from the snowy part of the Carpathian range, is 18 fathoms high and 9 broad; and the cave spreads out laterally, and descends to a point 50 fathoms below the entrance, where it is 26 fathoms in breadth, and of irregular height. Beyond this no one had at that time penetrated, on account of the unsafe footing, although many distant echoes

---

[98] The *Caves of Szelicze* are mentioned in Murray's *Handbook of Southern Germany* (1858, p. 555), where the following account is given of them:—'During the winter a great quantity of ice accumulates in these caves, which is not entirely melted before the commencement of the ensuing winter. In the summer months they are consequently filled with vast masses of ice broken up into a thousand fantastic forms, and presenting by their lucidity a singular contrast to the sombre vaults and massive stalactites of the cavern.'

The *Drachenhöhle* (Murray, 1. c.p. 553), a series of caverns not far from Neusohl in Hungary, afford another instance of an ice-cave, one of the largest of them being said to be coated with a sheet of translucid ice, through which the stalactitic fretwork of the vault is seen to great advantage.

[99] Not far from Kaschau.

were returned by the farther recesses of the cave; indeed, to get even so far as this, much step-cutting was necessary.

When the external frost of winter comes on, the account proceeds, the effect in the cave is the same as if fires had been lighted there: the ice melts, and swarms of flies and bats and hares take refuge in the interior from the severity of the winter. As soon as spring arrives, the warmth of winter disappears from the interior, water exudes from the roof and is converted into ice, while the more abundant supplies which pour down on to the sandy floor are speedily frozen there. In the Dog-days, the frost is so intense that a small icicle becomes in one day a huge mass of ice; but a cool day promptly brings a thaw, and the cave is looked upon as a barometer, not merely feeling, but also presaging, the changes of weather. The people of the neighbourhood, when employed in field-work, arrange their labour so that the mid-day meal may be taken near the cave, when they either ice the water they have brought with them, or drink the melted ice, which they consider very good for the stomach. It had been calculated that 600 weekly carts would not be sufficient to keep the cavern free from ice. The ground above the cave is peculiarly rich in grass.

In explanation of these phenomena, Bell threw out the following suggestions, which need no comment. The earth being of itself cold and damp, the external heat of the atmosphere, by partially penetrating into the ground, drives in this native cold to the inner parts of the earth, and makes the cold there more dense. On the other hand, when the external air is cold, it draws forth towards the surface the heat there may be in the inner part of the earth, and thus makes caverns warm. In support and illustration of this view, he states that in the hotter parts of Hungary, when the people wish to cool their wine, they dig a hole 2 feet deep, and place in it the flagon of wine, and, after filling up the hole again, light a blazing fire upon the surface, which cools the wine as if the flagon had been laid in ice. He also suggests that possibly the cold winds from the Carpathians bring with them imperceptible particles of snow, which reach the water of the cave, and convert it into ice. Further, the rocks of the Carpathians abound in salts, nitre, alum, &c., which may, perhaps, mingle with such snowy particles, and produce the ordinary effect of the snow and salt in the artificial production of ice.

Townson[100] visited this cave half a century later, and concluded that Bell was in error with regard to the supposed winter thaw and summer frost, although he himself received information at Kaschau which corroborated the earlier account. He describes the approach to the village of Szilitze as leading by a by-road through a pleasant country of woods and hills, with much pasture-land, the cave lying a mile beyond the village, and displaying an entrance 100 feet broad, and 20 or 30 feet high, turned towards the north. The descent of the floor of the cave is rapid, and was covered with thin ice, at the time of his visit, for the last third of the way: from the roof at the farther end, where the cave is not so high as at the entrance, a congeries of icicles was seen to hang; and in a corner on the right, completely sheltered from the rays of the sun, there was a large mass of the same material. It was a fine forenoon in July, and all was in a state of thaw, the icicles dropping water, and the floor of ice covered with a thin layer of water; while the thermometer in all parts of the cave stood at zero of Réaumur's scale. The rock is compact unstratified limestone, in which so many of the famous caverns of the world are found.

THE CAVE OF YEERMALIK, IN KOONDOOZ[101]

In the year 1840, Captain Burslem, of the 13th Light Infantry, made an expedition from Cabul to the North-west, accompanied by Lieutenant Sturt of the Bengal Engineers, who was afterwards killed in the terrible pass where Lady Sale, whose daughter he had married, was shot through the arm.

After crossing the high and wild pass of Karakotul (10,500 feet), these travellers reached the romantic glen of the Doaub, which lies at the foot of the pass, and is surrounded on all sides by lofty mountains. Here they were hospitably entertained by Shah Pursund Khan, the chief of the small

---

[100] *Travels in Hungary*, 1797, pp. 317, &c.
[101] *A Peep into Toorkistan*; London, 1846; chapters x. and xi.

territory, and their curiosity was roused by the account given by an old moollah of a cavern seven miles off, which the Shah strongly advised them not to attempt to visit, for the Sheitan (the devil), whose ordinary place of abode it was, never allowed a stranger to return from its recesses. The moollah, however, scouted this idea, on the ground that it was much too cold for such an inhabitant; and the Shah eventually agreed to accompany them to the cave with a band of his followers.

As they rode through long and rich grass, following the course of a gentle stream, and tormented by swarms of forest flies, or blood-suckers, the Shah informed them that he had once endeavoured to explore the cave, and had already penetrated to a considerable distance, when he came upon the fresh prints of a naked foot, with an extraordinary impression by their side, which he suspected to be the foot of Sheitan himself, and so he beat a precipitate retreat. The moollah told them that there was a large number of skeletons in the cave, the remains of 700 men who took refuge there during the invasion of Genghis Khan, with their wives and families, and defended themselves so stoutly, that, after trying in vain the means by which the M'Leods were destroyed in barbarous times, and the opponents of French progress in Algeria in times less remote, the invader built them in with huge natural blocks of stone, and left them to die of hunger.

The entrance is half-way up a hill, and is 50 feet high, with about the same breadth. Not far from the entrance they found a passage between two jagged rocks, possibly the remains of Genghis Khan's fatal wall, so narrow that they had some difficulty in squeezing through; and then, before long, came to a drop of 16 feet, down which they were lowered by ropes made from the cotton turbans of the Shah and his attendants. Here they left two men to haul them up on their return, and bade farewell to the light of day. The narrow path led by the edge of a black abyss, sometimes over a flooring of smooth ice for a few feet, and widened gradually till they reached a damp and dripping hall, of dimensions so vast that the light of their torches did not enable them to form a conception of its size. In this hall they found hundreds of skeletons in a perfectly undisturbed state, one, for instance, still holding the skeletons of two infants in its bony arms, while some of the bodies had been preserved, and lay shrivelled like those at the Great St. Bernard. They were very much startled here by the discovery of the prints of a naked human foot, and by its side the distinct mark of the pointed heel of an Affghan boot,[102] precisely what had so thoroughly frightened the Shah twelve years before. The prints retained all the sharpness of outline which marks a recent impression, and led towards the farther recesses of the cave; but the Englishmen were called away from their investigation by the announcement that if they did not make haste, there would not be oil enough for lighting them to the ice-caves.

Proceeding through several low arches and smaller caves, they reached at length a vast hall, in the centre of which was[103] an enormous mass of clear ice, smooth and polished as a mirror, and in the form of a gigantic beehive, with its dome-shaped top just touching the long icicles which depended from the jagged surface of the rock. A small aperture led to the interior of this wonderful congelation, the walls of which were nearly 2 feet thick; the floor, sides, and roof were smooth and slippery, and their figures were reflected from floor to ceiling and from side to side in endless repetition. The inside of this chilly abode was divided into several compartments of every fantastic shape: in some the glittering icicles hung like curtains from the roof; in others, the vault was smooth as glass. Beautifully brilliant were the prismatic colours reflected from the varied surface of the ice, when the torches flashed suddenly upon them as they passed from cave to cave. Around, above, beneath, everything was of solid ice, and being unable to stand on account of its slippery nature, they slid, or rather glided, mysteriously along the glassy surface of this hall of spells. In one of the largest compartments the icicles had reached the floor, and gave the idea of pillars supporting the roof.

The cavern in which this marvellous mass of ice stood, branched off into numerous galleries, one of which led the party to a sloping platform of rapidly increasing steepness, where they were startled by the reappearance of the naked foot-prints, passing down the slope. The toes were spread out in a manner which showed that they belonged to someone who had been in the habit of going barefoot, and Captain Burslem took a torch and

---

[102] They were now in a country far removed from the Affghans, and hostile to that people.
[103] The remainder of this paragraph is in Captain Burslem's own words.

determined to trace the steps: a large stone, however, gave way under his weight; and this, sliding down at first, and then rolling and bounding on forever, raised such a tumult of noise and echoes that the natives with one accord cried 'Sheitan! Sheitan!' and fled precipitately, extinguishing all the lights in their fear; so that but for Sturt's torch the whole party must have been lost in the darkness. Shah Pursund Khan at once called a retreat, vowing that it was of no use to attempt to follow the footsteps, as it was well known that the cave extended to Cabul! The guides had now lost their small allowance of pluck, and wandered about despairingly for a long time before they could find their way back to the ice-cave, and thence to the foot of the rock where the two men and the turban-ladders had been left. As soon as they came in sight of this, their comrades above cried out to them that they must make all haste, for Sheitan himself had appeared an hour before, running along the ledge where they now were, and finally vanishing into the gloom beyond; an announcement which of course produced a stampede in the terrified party of natives. Five or six rushed to the spot where the turbans hung, and only an opportune fall of stones from above prevented their destroying the apparatus in their blind hurry to escape. The chief claimed the privilege of being drawn up first, and he and all his followers declared that nothing should ever tempt them to visit again the Cave of Yeermalik.[104]

The Surtshellir, in Iceland

The first account of this lava-cavern is given by Olafsen,[105] who visited it in 1750 and 1753. Ebenezer Henderson[106] explored it in 1815, and Captain Forbes gives some account of it in his recent book on Iceland.[107] It

---

[104] I am indebted for the knowledge of the existence of these caves to W.A. Sandford, Esq., F.G.S., who informed me that an account of them was to be found in a book of travels by an English officer. I am not aware that they have been visited on any other occasion than this.

[105] *Reise durch Island*, Copenhagen, 1744 (being a German translation from the original Danish), i. 128 sqq.

[106] Henderson's Iceland, ii. 189 sqq.

[107] Pp. 145 sqq.

is mentioned in some of the Sagas,[108] and appears to have been a refuge for robbers in the tenth century, and Sturla Sigvatson, with a large band of followers, spent some time here. The Landnama Saga derives the name Surtshellir from a huge giant called Surtur, who made his abode in the cave; but Olafsen believed that the name merely meant *black hole*, from *surtur* or *svartur*, and was due to the darkness of the cave and the colour of the lava: in accordance with this view, it is called *Hellerin Sortur*, or *black hole*, in some of the earlier writings. The common people are convinced that it is inhabited by ghosts; and Olafsen and his party were assured that they would be turned back by horrible noises, or else killed outright by the spirits of the cave: at any rate, their informants declared they would no more reach the inner parts of the cavern than they had reached the traditional green valley of Aradal, isolated in the midst of glaciers, with its wild population of descendants of the giants, which they had endeavoured to find some time before.[109]

The cave is in the form of a tunnel a mile or more in length, with innumerable ramifications, in the lava which has flowed from the Bald Yökul. It lies on the edge of the uninhabited waste called the Arnavatns-heidi, in a district described by Captain Forbes as distorted and devilish, a cast-iron sea of lava. The approach is through an open chasm, 20 to 40 feet in depth, and 50 feet broad, leading to the entrance of the cave, where the height is between 30 and 40 feet, and the breadth rather more than 50. Henderson found a large quantity of congealed snow at this entrance, and along pool of water resting on a floor of ice, which turned his party back and forced them to seek another entrance, where again they found snow piled up to a considerable height. Olafsen also mentions collections of snow under the various openings in the lava which forms the roof of the cave. The latter explorer discovered interesting signs of the early inhabitants of the Surtshellir, as, for instance, the common bedstead, built

[108] The Sturlunga, Landnama, and Holmveria Sagas.

[109] Two priests determined to solve the mystery of this unapproachable valley, the Aradal, or Thoris-thal, with its rich meadows and gigantic inhabitants, and made an expedition for this purpose in 1664. They reached a point where the glaciers fell off into a valley so deep that they could not see whether there were meadows at the bottom or not, and the slope was so rapid that it was impossible to descend.

of stones, 2-1/2 feet high, 36 feet long, and 14 feet broad, with a pathway down the middle, forming the only passage to the inner parts of the cave. The spaces enclosed by these stones were strewn with black sand, on which rough wool was probably laid by way of mattress. This could scarcely have been a bedstead in the time of the giants, for a total breadth of 14 feet, deducting for the pathway down the middle, will not give more than 6 feet for the layer of men on either side, unless indeed they lay parallel to the passage, and required a length of 36 feet. He also found an old wall, built with blocks of lava across one part of the cave, as if for defence, and a large circular heap of the bones of sheep and oxen, presumably the remains of many years of feasting. Captain Forbes scoffs at these bones, and suggests errant wild ponies as the depositors thereof.

Olafsen had found in his earlier visit that the way was stopped, far in the recesses of the cave, by a lake of water, which filled the tunnel to a depth of 3 feet or more, lying on ice; but in 1753 there was not more than a foot of water, through which they waded without much difficulty. The air soon became exceedingly cold and thick, and for some hundreds of paces they saw no light of day, till at length they reached a welcome opening in the roof. Beyond this, the air grew colder and more thick, and the walls were found to be sheeted with ice from roof to floor, or covered with broad and connected icicles. The ground also was a mass of ice, but an inch or two of fine brown earth lay upon it, which enabled them to keep their footing. This earth appeared to have been brought down by the water which filtered through the roof. 'The most wonderful thing,' Olafsen remarks, 'that we noticed here, was, that the stalactites of ice were set with regular figures of five and seven sides, joined together, and resembling those seen on the second stomach of ruminating animals. The condensed cold of the air must have imparted these figures to the ice; they were not external (merely?), but in the ice itself, which otherwise was clear and transparent.'

Henderson and his party appear to have had much more wading to do than Olafsen, walking in one instance through a long tract of water up to the knees. In the deeper recesses of the cave, apparently in the part where the earlier explorers had found the reticulated ice, they found the whole floor of the passage covered with thick ice, with so steep a dip that they sat down and slid forward by their own weight--a most undignified proceeding for a grave gentleman on a mission from the Bible Society. On holding their torches close to the floor, they saw down to a depth of 7 or 8 feet, the ice being as clear as crystal. 'The roof and sides of the cave were decorated with most superb icicles, crystallised in every possible form, many of which rivalled in minuteness the finest zeolites; while from the icy floor rose pillars of the same substance, assuming all the curious and phantastic shapes imaginable, mocking the proudest specimens of art, and counterfeiting many well-known objects of animated nature. Many of them were upwards of 4 feet high, generally sharpened at the extremity, and about 2 feet in thickness. A more brilliant scene perhaps never presented itself to the human eye, nor was it easy for us to divest ourselves of the idea that we actually beheld one of the fairy scenes depicted in Eastern fable. The light of the torches rendered it peculiarly enchanting.'

Captain Forbes found much ice on the floor, but he did not enjoy the cold and wet, and seems to have ascended by the last opening in the roof, mentioned by Olafsen, before reaching the cavern where the more beautiful parts of the ice-decoration were found by his predecessors. The two engravings of the interior of the cave given in his book are copied from the magnificent lithographs of Paul Gaimard,[110] but much of the effect has been lost in the process of copying.

Mr. Baring Gould mentions this cavern in his book on Iceland, and believes that its interest has been much overrated. He seems to have visited the cave, but makes no allusion to the existence of ice.[111]

Mr. E.T. Holland visited the Surtshellir in the course of his tour in Iceland, in 1861, and an account of his visit is given in the first volume of 'Peaks, Passes, and Glaciers.'[112] After following in Olafsen's steps for some time, the party reached a cave whose floor was composed of very clear ice, apparently of great thickness, for they could not see the lava beneath it. The walking on this smooth ice-floor Mr. Holland describes as being

---

[110] Voyage en Islande; Atlas Historique; t. ii., pl. 130-133.
[111] Iceland: its Scenes and Sagas: pp. 97, 98.
[112] Page 113.

delightful, the whole sloping considerably downwards. 'In five minutes,' he continues, 'we reached the most beautiful fairy grotto imaginable. From the crystal floor of ice rose up group after group of transparent icy pillars, while from the glittering roof most brilliant icy pendants hung down to meet them. Columns and arches of ice were ranged along the crystalline walls ... I never saw a more brilliant scene; and indeed it would be difficult to imagine anything more fairy-like. The pillars were many of them of great size, tapering to a point as they rose. The largest were at least 8 feet high, and 6 feet in circumference at their base. The stalactites were on an equally grand scale. Through this lovely ice-grotto we walked for nearly ten minutes.'

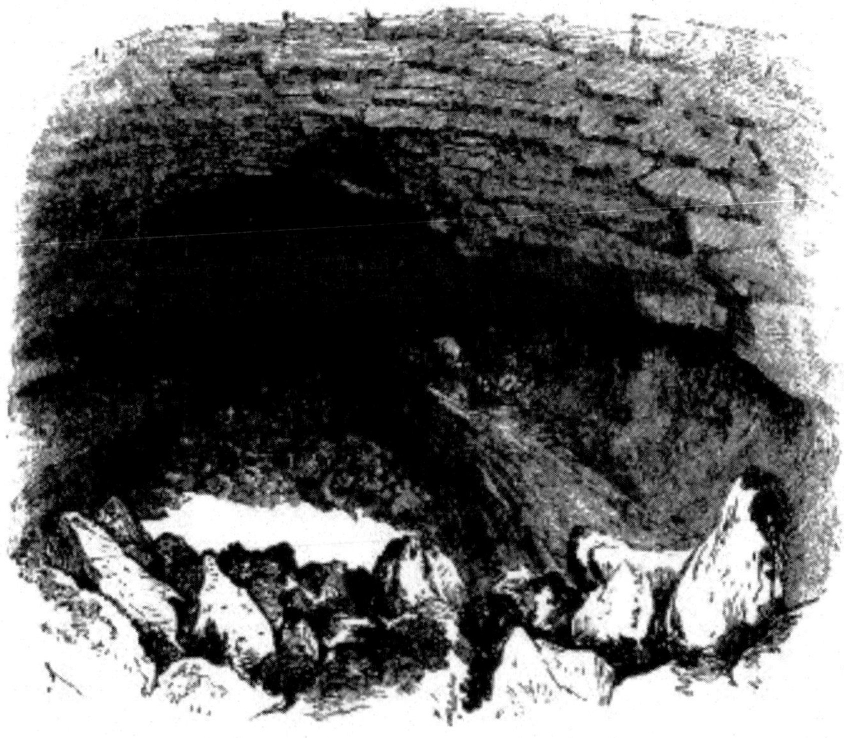

Figure 13. Ice-Cave in the Surtshellir.

The temperature of the caves, Mr. Holland states in a note, was from 8° to 10° C. (46·4° to 50° F.), that of the air outside being 53·6° F.

The Gypsum Cave of Illetzkaya-Zastchita, in the Steppes of the Kirghis, South of Orenburg

The district in which this cavern occurs is a small green oasis on the undulating steppe, lying on a vast bed of rock-salt, which extends over an area of two versts in length, and a mile in breadth, with a thickness of more than 100 feet. When the thin cover of red sand and marl is removed, the white salt is exposed, and is found to be so free from all stain, or admixture of other material, excepting sometimes minute filaments of gypsum, that it is pounded at once for use, without any cleansing or recrystallising process.

In the immediate neighbourhood of Illetzkaya-Zastchita there are two or three gypseous hillocks, and a cavern in one of these is used by the inhabitants as a cellar, having been artificially enlarged for that purpose. Sir Roderick Murchison and his colleagues visited this cavern on a hot day in August, with the thermometer at 90° in the shade, in the course of their travels under the patronage of the late Emperor of Russia.[113] They found the hillock to be an irregular cone 150 feet in height; the entrance was by a frail door, on a level with the village street, and fully exposed to the rays of the sun; and yet, when the door was opened, so piercing a current of cold air poured forth, that they were glad to beat a retreat for a while; and on eventually exploring farther, they found the quass and provisions, stored in the cave, half-frozen within three or four paces of the door. The chasm soon opened out into a natural vault from 12 to 15 feet high, 10 or 12 paces long, and 7 or 8 in width, which seemed to have numerous small ramifications into the impending mound of gypsum and marl. The roof of this inner cavern was hung with undripping solid icicles, and the floor was a conglomerate of ice and frozen earth. They were assured that the cold is always greatest within when the external air is hottest and driest, and that the ice gradually disappears as winter approaches, and vanishes when the snow comes. The peasants were unanimous in these statements, and

[113] Russia and the Ural Mountains, i. 186, sqq.

asserted that they could sleep in the cave without sheepskins in the depth of winter.

Sir Roderick Murchison and his friends were at first inclined to explain these phenomena by supposing that the chief fissure communicated with some surface of rock-salt, 'the saliferous vapours of which might be so rapidly evaporated or changed in escaping to an intensely hot and dry atmosphere as to produce ice and snow.' But Sir John Herschel, to whom they applied for assistance, rejected the evaporation theory, and suggested that the external summer wave of heat might possibly only reach the cave at Christmas, being delayed six months in its passage through the rock; the cold of winter, in the same manner, arriving at midsummer. To this the explorers objected, that the mound contained many caves, but' only in this particular fissure was any ice found. Dr. Robinson, astronomer at Armagh, endeavoured to explain the matter by referring to De Saussure's explanation of the phenomena of cold caves in Italy and elsewhere; but this, too, was considered unsatisfactory. At length, Professor Wheatstone referred them to the memoir by Professor Pictet, in the Bibliothèque Universelle of Geneva, where that savant improves upon De Saussure's theory, and applies it in its new form to the case of caves containing permanent ice, in tracts whose mean cold is above the freezing point. This they seem to have accepted, adding that the climatological circumstances of Orenburg--a wet spring, caused by the melting of the abundant snows, followed by a summer of intense and dry Asiatic heat--must be particularly favourable for the working out of the theory, and must also act powerfully in producing the refrigerating effects of evaporation.[114]

The traveller Pallas visited Illetzkaya in July 1769, and describes this gypseous hillock.[115] In his time the entrance by the side of the hill was unknown, as also was the existence of ice in the cavern. He saw at the top

---

[114] See the Papers read before the Geological Society of London, on March 9, 1842, by Sir John Herschel and Sir E. Murchison, the substance of which has been given above.
See also the *Edinburgh Philosophical Journal* for 1843 (xxxv. 191), for an attempt by Dr. Hope to explain the phenomena of this cave by a reference to the slow penetration of the winter and summer waves of cold and heat. Dr. Hope believes that, although the external changes do not travel to any great depth, they reach far enough to communicate with some of the fissures leading to the cave.

[115] *Voyages* (French translation); Paris, 1788; i. 364.

of the Kraoul-naï-Gora, or Watch-mountain, as it was called, a fissure which had once formed a large cavern, into which the Kirghis were in the habit of throwing furs and other materials as religious offerings. Although the cave had since fallen in, they still kept up a part of the ceremony, marching solemnly round the base of the hill once a year, and bathing in the neighbouring water. In earlier times, a man had descended through the fissure by means of cords, and found the cold within insupportable, having very probably reached the present ice-cave.

Pallas describes many caves in various parts of Russia, but never seems to hint at the existence of ice in them, though he specially mentions their extreme cold. Some of these occurred in gypsum, and some in limestone; and the gypseous caves showed universally a very low temperature, though still far above the freezing-point.[116] Thus in the dark cavern of Barnoukova,[117] on the Piana, in a rock of gypsum, while the thermometer in the shade stood at 75°•2, the temperatures at various points in the cave were,--at the entrance 59°•36, 25 feet from the entrance 46°•4, and in the coldest part 42°•8. This cold he describes as insupportable. The temperature of the water which had accumulated in the coldest parts of the cave was 48°•8, considerably higher than the surrounding atmosphere; from which Pallas concluded that the cold of gypsum-caves is due to the acid vapours which are generally observed in grottoes of this description. In May 1770, he found snow on the sloping entrance to the cavern of Loeklé, in the neighbourhood of the Oufa; but the air of the interior was not colder than was to be expected in a deep cave.

Sir R. Murchison wrote to Russia for further information with respect to this cave in January 1865, and again in the beginning of April, addressing his second enquiry to the Secretary of the Imperial Academy. In reply, the Secretary says that he is not aware that any thermometric observations have been made in the cavern. He encloses a short statement

---

[116] In the gypsum to the NE. of Kungur, on the banks of the Iren, there is a cave containing ice. Four of its chambers have ice, in one of which a stalagmite of ice rises almost to the roof. The farthest chamber, 625 fathoms from the entrance, contains a lake of water which stretches away out of sight under the low roof. (*Taschenbuch für die gesammte Mineralogie*; Leonhard, 1826; B. 2, S. 425. Published as *Zeitschrift für Mineralogie*.)

[117] Pallas, *Voyages*, i. 84.

by M. Helmersen, one of the members of the Academy, to the following effect:--About 50 versts SE. of Miask, in the chain of the Ural, is a copper mine, called Kirobinskoy, which was abandoned more than fifty years ago. On the 7th July, 1826, M. Helmersen found a thick wainscoting of ice on the sides and roof and floor of the horizontal gallery, within 10 feet of the entrance. He was assured that this ice never melts, and that its thickness is greater in summer than in winter. M. Helmersen adds, that to the best of his belief no one has investigated the cavern of Illetzkaya Zastchita since Sir R. Murchison's visit.

THE ICE-CAVERN OF THE PEAK OF TENERIFFE[118]

This cave is at a height of 11,040 feet above the sea, and is therefore not far below the snow-line of the latitudes of the Canary Isles. The entrance is by a hole 3 or 4 feet square, in the roof of the cave, which may be about 20 feet from the floor. The peasants who convey snow and ice from the cave to the lower regions, enter by means of knotted ropes; but Professor Smyth had caused his ship's carpenter to prepare a stout ladder, by which photographic instruments and a lady were taken down.

On alighting on a heap of stones at the bottom, the party found themselves surrounded by a sloping wall of snow, 3 feet high, and 7 or 8 feet broad, the basin in which they stood being formed in the snow by the vertical rays of the sun, and by the dropping of water from the edges of the hole.[119] Beyond this ring-fence, large surfaces of water stretched away into the farther recesses of the cave, resting on a layer of ice, which appeared to be generally about 2 feet thick. At one of the deeper ends of the cave, water dropped continually from the crevices of the roof; a fact which Professor Smyth attributed to the slow advance of the summer wave of heat through the superincumbent rock, which was only now reaching the

[118] *Teneriffe*, by Professor Smyth, ch. viii., and Humboldt, *Voyage aux Régions Équinoctiales*; Paris, 1814; i. 124.

[119] They afterwards discovered smoke issuing from the centre of this patch of stones; so that volcanic heat may possibly have had something to do with the disappearance of the snow.

inner recesses of the loose lava, and liquefying the results of the past winter. There would seem to be immense infiltration of meteoric water on the Peak; for, notwithstanding the great depth of rain which falls annually in a liquid or congealed form, the sides of the mountain are not scored with the lines of water-torrents.

Though occurring in lava, this cavern is quite different from lava-tunnels, such as the Surtshellir, which are recognised formations, produced by the cooling of the terminal surface-crust of the stream of lava, and the subsequent bursting forth of the molten stream within. This, on the contrary, proved to be a smooth dome-shaped cave, running off into three contracting lobes or tunnels which might be respectively 70, 50, and 40 feet long, and were all filled to a certain depth with water: in the smoothness of the interior surfaces, Professor Smyth believed that he detected the action of highly elastic gases on a plastic material.

The astronomer takes exception to the term 'underground glacier'[120] which had been applied to this cavern. He represents that the mountain is abundantly covered each winter with snow, in the neighbourhood of the ice-cave, which is nearly within the snow-line, and the stores of snow thus accumulated in the cave have no great difficulty in resisting the effects of summer heat, since all radiation is cut off by the roof of rocks. The importance of this protection may be understood from the fact that in the middle of July the thermometer at this altitude gave 130° in the sun, but fell to 47° when relieved from the heat due to radiation. At the time of this observation, there were still patches of snow lying on the mountain-side, exposed to the full power of direct radiation; and, therefore, there is not anything very surprising in the permanence of snow under such favourable circumstances as are developed in the cave. Mr. Airy, a few summers ago, found the rooms of the Casa Inglese, on Mount Etna, half filled with snow, which had drifted in by an open door, and had been preserved from solar radiation by the thick roof.[121]

---

[120] 'Ce petit glacier souterrain,' Humboldt, l.c.

[121] See p. 272 for an account of the underground glacier in the neighbourhood of the Casa Inglese.

Humboldt remarks, that the mean temperature of the region in which the Cueva del Hielo (ice-cave) occurs, is not below 3° C. (37·4° F.), but so much snow and ice are stored up in the winter that the utmost efforts of the summer heat cannot melt it all. He adds, that the existence of permanent snow in holes or caves must depend more upon the amount of winter snow, and the freedom from hot winds, than on the absolute elevation of the locality.

The natives of Teneriffe are men of faith. They have large belief in the existence and intercommunication of numerous vast caverns in the Peak, one of which, on the north coast, is said to communicate with the ice-cavern, notwithstanding 8 miles of horizontal distance, and 11,000 feet of vertical depth. The truth of this particular article of their creed has been recently tested by several worthy and reverend hidalgos, who drove a dog into the entrance of the cavern on the sea-coast, in the belief that he would eventually come to light again in the ice-cave: he was accordingly found lying there some days after, greatly fatigued and emaciated, having in the interval accomplished the 11,000 feet of subterranean climbing. How he could enter, from below, a water-logged cave, does not appear to have been explained.

Chapter 16

BRIEF NOTICES OF OTHER ICE-CAVES[122]

On the Brandstein in Styria, in the district of Gems, there is an ice-hole closely resembling some of the glacières of the Jura. It is described by Sartori,[123] as lying in a much-fissured region, reached after four hours of steep ascent from the neighbouring village, through a forest of fir. Some of the fissures contain water and some snow, while others are apparently unfathomable. From one of the largest of these, a strong and cold current blows in summer, and in this fissure is the ice-hole. Sartori found *crimpons* necessary for descending the frozen snow which led from the entrance to the floor of the cave, where he discovered pillars and capitals and pyramids of ice of every possible shape and variety, as if the cave had contained the ruins of a Gothic church, or a fairy palace. At the farther end, after passing large cascades of ice, his party reached a dark grey hole, which lighted up into blue and green under the influence of the torches; they could not discover the termination of this hole, and the stones which they rolled down into it seemed to go on for ever. The greatest height of the cave is about 36 feet, and its length 192 feet, with a maximum breadth of 126 feet. Towards the end of autumn, the temperature of the ice-hole rises so much,

---

[122] Several of these caves are referred to by Reich, Beobachtungen über die Temperatur des Gesteins in verschiedenen Tiefen in den Gruben des Sächsischen Erzgebirges; Freiberg, 1834.

[123] Naturwunder des Oesterr. Kaiserthums, iii. 40.

that the glacial decorations disappear, and various wild animals are driven by the cold of winter to take shelter in the comparative warmth of the cave. The elevation of the district in which this ice-hole occurs is about 1,800 German feet above the sea.

In Upper Styria, where the Frauenmauer overlooks the basin in which the mining town of Eisenerz is situated, an ice-cave has been explored, and a description of it has been given by certain members of the Austrian Alpine Club.[124] The Brandstein is spoken of as one of the peaks in the immediate neighbourhood; and as the cave previously described is stated by Sartori to be on the Brandstein, that district would seem to be rich in glacières. The cavern is most easily explored from Eisenerz, and on that side the entrance is 4,539 Vienna feet above the sea. Its other outlet, in the Tragöss valley, is 300 feet higher. The total length of the cave is 2,040 Vienna feet. After passing the entrance, which is an archway from 12 to 18 feet high, the main course of the cave is soon left, and a branch is followed which leads to the *Eis-kammer*. This ice-chamber consists of a grotto from 30 to 40 fathoms long, decked with ice-crystals, pillars of ice, and cascades of the same material, the floor being composed of ice as smooth as glass. In the summer, pleasure-parties assemble in the cave and amuse themselves with the game of *Eisschiessen*, so popular in Upper Styria as a winter diversion. The hotter the summer, the more ice is found in the Eiskammer, and the general belief is that it all disappears in winter.

The cave proper, which assumes stupendous dimensions in its long course, shows no ice. It seems to be formed in the Muschelkalk of the Trias formation, and so far no limestone stalactites have been discovered. It has not, however, as yet been fully explored. The editor of the proceedings of the Austrian Alpine Club gives a reference to Scheiner, '*Ausflug nach der Höhle der Frauenmauer,' (Steiermarkische Zeitschrift, neue Folge*, i. 2, 1834, p. 3.)

---

[124] *Mittheil. des Oesterr. Alpen-Vereins*, ii. 441. I am indebted to G.C. Churchill, Esq., one of the authors of the well-known book on the Dolomite Mountains, for my knowledge of the existence of this cave, and of the Kolowrathöhle.

At Latzenberg, near Weissenstein in Carniola, there is another ice-cave, described by Rosenmüller.[125] It is entered by a long dark passage in which are pillars of ice arranged like the pipes of an organ, varying from the thickness of a man's body to the size of a straw. All these are said to melt in winter. Farther on are two other passages, one of which passes upwards over *Stufe*, and is coated in summer with ice; the other has not been explored.

Near Glaneck in the Untersberg, not far from Salzburg, is a cave called the Kolowrathöhle, of which a description is given by Gümbel in his great geological work on the Bavarian Alps.[126] It is a spacious cavern, opening in a steep wall of rock above the *Rositenschlucht* between the Platten and *Dachstein-kalk.*[127] An ice-current rushes from within, and ice is found on the threshold, becoming more prevalent in the farther recesses of the cave. The lower parts are tolerably roomy, and masses of ice of various shapes are found piled one upon another, lighting up with magical effect when torches are brought to bear upon them. Gümbel believes that the cold currents which stream into the cave from the numerous fissures in its walls are the cause of the ice; and though this is the only known ice-cave far and near, he imagines that the icy-currents which are frequently met with in that district, and in the *Hochgebirge*, would be found to proceed in reality from like caves, if the fissures from which they blow could be penetrated.

Behrens[128] describes two ice-caves near Questenberg, in the county of Stollberg, on the Harz mountains. They both occur in limestone, and are known as the Great and Little Ice-holes. The one is close to the village of Questenberg, and consists of a chasm several fathoms deep, so cold that in summer the water trickling down its edges is frozen into long icicles. The opening is large and faces due south, and yet the hotter the day the more ice is found; whereas in winter a warm steam comes out, as if from a stove.

[125] Beschreibung merkwürdiger Höhlen, ii. 283.

[126] Geognostísche Reschreibung des bayerischen Alpengebirges; Gotha, 1861.

[127] These constitute the upper bone bed and Dachstein limestone beds of the uppermost part of the Trias formation.

[128] Hereynia Curiosa, cap. v. The same account is given in Behren's Natural History of the Harz Forest, of which an English translation was published in 1730.

The other cave is farther into the mountain; it is spacious and light, and very cold in summer.

In Gehler's *Physik. Wörterbuch* (Art Höhle), a small hole is mentioned near Dôle, which is said to be remarkable for the large and curiously-shaped icicles found there; but no sufficient account of it seems to have been given.

An ice-hole is also spoken of in the same article, which occurs on the east side of the town of Vesoul.[129] The hole is described as being small, with a little rivulet of water: this water, and also that which trickles down the walls of the cave, is converted into ice, and so much is formed on a cold day that it requires eight warm days to melt it. Gollut, in his description of the *fré-puits* of Vesoul,[130] observes that the remarkable pit known by that name was so cold, that in his time it had never been fully explored. Gehler's expression, however, 'a small hole,' cannot possibly apply to the *fré-puits*; so that these would seem to be two different examples of cold caves near Vesoul.

There is an interesting account in Poggendorff's Annalen[131] of a visit made by Professor A. Pleischl to a mountain in the circle of Leitmeritz, where ice is found in summer under very curious circumstances. The mountain is called Pleschiwetz, and lies above Kameik, in Bohemia, not far from the town of Leitmeritz. On the 24th of June in each year, large numbers of pilgrims assemble at the romantic chapel of S. John the Baptist in the Wilderness; and it is a part of their occupation to search for ice under the basaltic rocks, and carry it home wrapped in moss, as a proof that they have really made the pilgrimage. Professor Pleischl visited this district at the end of May 1834. The weather was hot for the season, as had been the case in April also, and there had been very little snow in the winter. A path leads from the chapel of S. John through the woods which deck the Pleschiwetz, and then over a small plain to the foot of the basaltic rocks. Here the mountain slopes away very steeply to the south, and the slope is thickly strewn with basaltic *débris*. From east to west this slope measures

---

[129] See also Muncke, *Handbuch der Naturlehre*, iii. 277; Heidelberg, 1830.

[130] See page 58. The more modern spelling is *frais-puits*.

[131] liv. 292.

about 40 fathoms, and its length is about 70 fathoms. It is surrounded on both sides and at the foot by trees and shrubs. The sun burned so directly on to the *débris*, that the basaltic blocks were in some cases too hot to be touched by the naked hand.

Professor Pleischl spent three hours of the early afternoon on this spot. The upper surface of the basaltic blocks had a temperature of at least 122° F. The presence of an icy current was detected by inserting the hand into the lower crevices; and on removing the loose stones to a depth of 1-1/2 or 2 feet, ice was found in considerable quantities. On the 27th of August, he proceeded to make a further investigation of this phenomenon; but he found the temperature of the blocks only 106° F., and in the crevices, at a depth of 2 or 3 feet, the lowest temperature reached was 38°·75 F. The external temperature in the shade was at the same time 83° F.

A third visit, in January 1835, gave no results; but on January 21, 1838, the Professor succeeded in determining some very remarkable facts. A depression in the sloping plain is called, *par excellence*, the ice-hole; and this is surrounded by firs and birches, which grow within three or four fathoms of the edge of the hole, so that the rays of the sun do not reach the hole in winter. Fresh snow lay on these trees; and there was nowhere any sign of melted snow, or of the formation of icicles. The basaltic *débris*, in which ice had been found in the summer, covers here a space of 5 fathoms long by 3 or 4 broad, immediately at the foot of a steep basaltic precipice. At eleven in the morning the temperature was 14° F. in the shade; and snow lay all round the ice-hole, to a thickness of 1-1/2 or 2 feet. The snow which covered the *débris* was pierced by holes, which could not have been caused by the sun, for its rays did not penetrate the trees; and, indeed, no sun had been visible for some days. These holes were generally turned towards the north, and were like chimneys. On investigation, it was found that icicles hung down into them, showing, of course, past or present thaw, and within the cavities no ice was found. The thermometer gave here from 27°·5 F. to 25°·15 F.; but in the crevices, into which the thermometer could not be pushed, the hand discovered a warm air. The moss drawn from these crevices was found to be steeped in unfrozen water, and it froze promptly when brought into the outer air.

The party afterwards climbed up the precipitous basalt, and reached, at 3 P.M., a level covered with large blocks of the same material, where the thermometer was slightly under 12° F. in the shade. The blocks were for the most part stripped of snow, and in some cases thin shields of ice were observed standing out two or three inches from them, forming hollow chambers, in which an agreeable warmth was found. These shields were invariably on the south side of the stones, the north side being free from ice and snow alike. In some places vapours were seen to rise. The thermometer gave 41° F. at a depth of six inches among the stones, though the external temperature, as has been said, was 12° F. For eight days previously, the thermometer had been always far below the freezing point, and on the 17th (four days before) had been 13° below zero (F.). On the 19th and 20th heavy snow had fallen. All these facts seem to show that the warmth which had caused the chimneys in the snow over the ice-holes, and the heated vapours on the higher parts of the mountains, proceeded from within, and not from without.

The people of the district assured Professor Pleischl that the hotter the summer, the more ice is formed; and that it disappears when the nights become long and the days short. Dr. Weiss, for six years head of the Gymnasium of Leitmeritz, stated that when one of the holes was emptied of ice in the summer, it filled again in a few days. The explanation given by the Professor of this phenomenon is, that the blocks of basalt, that being an excellent conductor of heat, pass so much warmth through to their under surfaces--which form the roof of small chambers filled with a spongy mass of decaying leaves--that the rapid evaporation thereby caused produces the cold air and the ice. He omits to explain why there should be anything exceptional in the winter phenomenon of the crevices among the stones.

There are two other places in Bohemia where ice is found in summer. One is on the Steinberg, in the county of Konaged;[132] it is a small basin, surrounded by trees, where, in the middle of summer, lumps of ice are found under basaltic *débris*. This ice is only formed, according to Sommer, in the hottest part of the year. The other is on the Zinkenstein, one of the

[132] Described by Schaller, *Leitmeritzer Kreis*, p. 271, and by Sommer, in the same publication, p. 331. I have not been able to procure this book.

highest points of the Vierzehnberg, in the circle of Leitmeritz. It is described by Sommer[133] as a cleft, five fathoms deep, in the basaltic rock, where ice is found in the hottest seasons. Professor Pleischl put this assertion to the test by visiting the spot in the end of August, when he found no signs of ice.

Another writer in Poggendorff[134] describes a somewhat similar appearance on the Saalberg. Here ice is found on the surface from June to the middle of August; and that, too, with a west exposure and in moderate shade. In July, the ice was so abundant that it could be seen from some distance: it was half a foot thick, and yielded neither to sun nor rain. In the middle of August there was no ice on the surface; but when the loose *débris* was removed, the most beautiful ice appeared, and at a little depth all was frozen as hard as if it had been the depth of winter.[135] The people who work in the neighbourhood declare that the place remains open, and free from ice or snow, in the greatest cold, and that no ice begins to form till the month of June. When the writer of the account in Poggendorff visited the ice-hole, the peasants were in the habit of carrying large masses of ice down to their houses, through a temperature of 81° F.

Reich[136] gives a detailed and valuable account of the prevalence of subterranean ice on the Sauberg, a hill which forms one side of a ravine near Ehrenfriedersdorf. The surface is about 2,000 feet above the sea, and its mean temperature, as determined by many careful observations, about 45° F. There are several tin-mines in this district, and the extended observations made by the authorities establish the curious fact that the mean temperature is considerably lower beneath than at the surface. For instance, in the S. Christoph pit, it is found that the mean temperature, at 15 fathoms below the surface, is only slightly above 42° F.; while at the

[133] *Böhmens Topogr.*, i. 339. This reference is given by Professor Pleischl.

[134] *Annalen*, lxxxi. 579.

[135] I was told, in 1864, by a chamois-hunter of Les Plans, a valley two hours above Bex, that some years before he was cutting a wood-road through the forest early in September, when, at a depth of 6 inches below the surface, he found the ground frozen hard. We visited the place together, but could find no ice. The whole ground was composed of a mass of loose round stones, with a covering of earth and moss, and the air in the interstices was peculiarly cold and dry.

[136] *Beobachtungen*, &c. (see note on p. 258), 181.

Morgenröther cross-cut the same mean temperature is found at a depth of 46 fathoms. The annual change of temperature is very small in these mines, and the maximum and minimum are reached very late; so that, if a point could be found with a mean temperature of 32° F., ice would increase there up to June or even July, and then diminish until December or January; in which case the phenomenon so often said to be observed in connection with subterranean ice--the melting in winter and forming in summer--would really be presented.

The ice on the Sauberg is frequently found to commence at a depth of 3 or 4 fathoms, and in the years 1811 and 1813 it extended to 24 fathoms below the surface: this depth, however, was exceptionally great, and as a rule the limit is reached at about 14 fathoms.[137] The ice is usually not very firm, and can be broken by stout blows with a stick; but between the years 1790 and 1800, when it was found at a depth of from 3 to 9 fathoms, it was so hard that blasting became necessary, and at that time the miners were with difficulty protected from the effects of the severe cold. The greatest quantity of ice is found in the interstices of the rubbish-beds of old workings, and here it assumes a crystalline form, the rocks being covered with a 'fibrous' structure, arranged perpendicularly to their surface.

Reich reports the universal presence of cold currents of air in these shafts and mines, and, in consequence, takes the opportunity of contradicting a statement in Horner's *Physik. Wörterbuch,*[138] that the absence of all current of air is essential to the formation of subterranean ice. He quotes the case of the cheese-caves of Roquefort as a further confirmation of his own observations with regard to the connection between ice in caves and cold currents of air; but of the many accounts which I have met with of the curious caves referred to, both in books and from the lips of those who have visited them, not one has made any mention of ice.[139] He states, too, that when the strength of the current is

---

[137] Reich found the temperature of the ice to be 31·982° F., that of the air in the immediate vicinity 34·025°, and the rock, at a little distance, 32·765°.

[138] iii. 150.

[139] See many careful descriptions of these caves in the Annales de Chimie; also, an account by Professor Ansted, in his Science, Scenery, and Art, p. 29. M. Chaptal (Ann. de Chimie, iv. 34) found the lowest temperature of the currents of cold air to be 36°·5 F.; but M. Girou de

diminished, its temperature is increased; a fact which all observations of the cold currents in caves, especially those made with so much care by M. Saussure, abundantly establish.

In the way of explanation, Reich mentions the possibility of rocks of peculiar formation possessing actually a low degree of temperature;[140] but he rejects this suggestion, preferring to believe that in some cases the cold resulting from evaporation is the cause of ice, and in others the greater specific gravity of cold as compared with warmer air.

In the *Bulletin des Sciences Naturelles*,[141] it is stated that a large quantity of ice is found in one of the recesses of the grotto of Antiparos--a fact which I have not seen mentioned elsewhere. After penetrating a long way through difficult fissures, a square chamber is at length reached, measuring 300 feet in length and breadth, with a height of about 80 feet. The walls and roof and floor are beautifully decorated with ice, and reflect all the colours of the rainbow. There are groups of pyramidal and round columns, and in some parts of the cave screens or curtains of ice 10 or 12 feet broad hang down to the floor.

In a later volume of the same periodical,[142] there is a description of a hill in Virginia where ice is found in summer. This hill lies near the road between Winchester and Romney, on the North River, latitude 39° N. One

---

Buzareingues (Ann. de Chimie et de Phys., xlv. 362) found that with a strong north wind, the temperature of the external air being 55°·4 F., the coldest current gave 35°·6 F.; with less external wind, still blowing from the north, the external air lost half a degree centigrade of heat, while the current in the cave rose to 38°·75 F. The cellars in which the famous cheese of Roquefort is ripened are not subterranean, but are buildings joined on to the rock at the mouths of the fissures whence the currents proceed. They are so valuable, that one, which cost 12,000 francs in construction, sold for 215,000 francs. The cheese of this district has had a great reputation from very early times. Pliny (Hist. Nat. xi. 97) mentions, with commendation, the cheeses of Lesura (M. Lozère or Losère) and Gabalum (Gevaudan, Javoux). The idolaters of Gevaudan offered cheeses to demons by throwing them into a lake on the Mons Helanus (Laz des Helles?) and it was not till the year 550 that S. Hilary, Bishop of Mende, succeeded in putting a stop to this practice.

[140] It would seem from his own account of the Sauberg, and from the description given above of the presence of ice among the rocky *débris*, as well as from the account on this page of ice in Virginia, that a formation of loose stones is favourable to the existence of a low degree of temperature. See also the note on p. 263, with respect to the loose stones near Les Plans. Forchhammer found, on the Faroë Islands, that springs which rise from loose stones are invariably colder than those which proceed from more solid rock at the same elevation, as indeed might have been expected.

[141] xvii. 337. The account is taken from a Dutch journal.

[142] xix. p. 124.

side of the hill is entirely composed of loose stones from ten to twenty pounds in weight, and under these the ice is found, although their upper surface is exposed to the full sun from 9 or 10 A.M. till sunset. In all seasons there is an abundance of icc. A writer in the 'London and Paris Observer'[143] visited the spot on the 4th of July, after a time of stifling heat, and in ten minutes he found more ice than the whole party could have carried away. He did not explore any farther than the foot of the hill; but the neighbours, who used the ice regularly in summer, assured him that it was to be found high up also. A constant and strong current issued from the crevices, stronger and infinitely colder than the current in the famous 'blowing cave' of Virginia. A man had built a store-room for meat within the influence of one of these currents, and hard dry icicles were seen hanging from the wooden supports inside: the flies, too, which had been attracted by the meat, were found frozen on to the stones. This is not the only district where ice is found within temperate latitudes in North America. In Professor Silliman's 'American Journal of Science,'[144] in a sketch of the geology of the township of Salisbury, Con. (latitude 43° N.), 'natural ice-houses' are mentioned. These consist of chasms of considerable extent in the mica-state, where ice and snow remain during the greater part of the year. The principal of these chasms lies in the east part of the town, and is several hundred feet long, sixty feet deep, and about forty wide. The slate is of a very compact kind; and the walls are perpendicular, and correspond with much exactness. At the bottom is a cold spring, and a cave of considerable extent, in which it is probable that the ice lies--for the writer does not specify the position in which it is found. The chasm is a favourite retreat in summer, and is called the Wolf-hollow, from its having formerly been a famous haunt for wolves.

Similar receptacles for summer-ice are found in several places in North America. In the forty-ninth volume of the *Sitzungsberichte der Kaiserl. Akademie in Wien* (1te. Abth.), a list of references to various ice-holes is appended to a paper by Dr. Boué on the geology of Servia. Many of the passages referred to have nothing to do with ice-caves, as, for instance, the

---

[143] October 11, 1829.
[144] viii. 254.

sections of De Saussure's book describing his observations of 'cold caves', or the account of the mass of ice and snow from which the river Jumna springs, for which Dr. Boué refers to the 'Philosophical Magazine' for November 1823, meaning, in fact, the 'London Magazine'. The 'Description des Glacières' of M. Bourrit is also given as a part of the literature on ice-caves; whereas (see the account of the Glacière of Montarquis, in the Valley of Reposoir) by 'glacière' M. Bourrit meant only a locality where ice is to be found, or a glacier district. Dr. Boué, however, gives some references to the 'American Journal of Science' which it is possible to make out by a careful search in the neighbourhood of the volume and page he mentions. In vol. iv. (1822,--Dr. Boué says 1821) there is an account by the editor[145] of a natural ice-house in the township of Meriden, Con., between Hartford and Newhaven, at an elevation of not more than 200 feet above the level of the sea. The ice is found in a narrow defile, which is hemmed in by perpendicular sides of trap-rock, and displays a perfect chaos of fallen blocks of stone. The defile is so narrow, that the sun's rays only reach it for an hour in the course of the day; and even the trees and rocks, and beds of leaves, protect the ice from any very material damage. Dr. Silliman visited this defile on the 23rd July, 1821,[146] with Dr. Isaac Hough, the keeper of a neighbouring inn, and found that the ice was only partially visible, in consequence of the large collection of leaves which lay on it: they sent a boy down with a hatchet, and he brought up some large firm masses, one of which, weighing several pounds, they carried twenty miles to Newhaven, where it did not entirely disappear till the morning of the third day. Seven miles from Newhaven, in the township of Branford, there is a similar collection of ice. In both of these cases, the ice is mixed with a considerable quantity of leaves and dirt.

In the same volume (p. 331,--Dr. Boué says p. 33), two accounts are given of a natural ice-house near the summit of a hill in the neighbourhood of Williamstown (Mass.). In the next volume there is a further account of it by Professor Dewey, stating that since the trees in the neighbourhood had

---

[145] Pp. 174-6.
[146] Thermometer about 85° F.

been cut, the snow and ice had disappeared each year about the first of August.

In vol. xlvi. (p. 331) an ice mountain in Wallingford, Rutland County (Vt.), is described, which is ordinarily known in the neighbourhood as the ice-bed. An area of thirty or fifty acres of ground is covered with massive *débris* of grey quartz from the mountains which overhang it; and here--especially in a deep ravine into which many of the falling blocks of stone have penetrated--ice is found in large quantities. It appears to be formed during the melting of the snow in February, March, and April, and vanishes in the course of the summer, in hot years as early as the last days of June.

These descriptions call to mind the Glacière of Arc-sous-Cicon, in which many of the features of the American ice-caves are reproduced. An American photograph is current in this country, in the form of a stereoscopic slide, representing an ice-cave in the White Mountains, New Hampshire; but it is only a winter cave, and in no way resembles any of the glacières I have seen. It is merely a collection of long and slender icicles, with beds of ice formed upon stones and trunks of trees on the ground; nothing more, in fact, than is to be seen in any tolerably severe winter in the neighbourhood of a cascade in a sheltered Scotch burn.

The 'American Journal of Science' (xxxvi. 184) gives a curious instance of a freezing-well near the village of Owego, three-quarters of a mile from the Susquehanna river. The depth of the well is 77 feet, and for four or five months in the year the surface of the water is frozen so hard as to render the well useless. Large masses of ice have been found in it late in July. A thermometer, which stood at 68° in the sun, fell to 30° in fifteen minutes at the bottom of the well; and the men who made the well were forced to put on thick clothing in June, and even so could not work for more than two hours at a time. No other well in that neighbourhood presents the same phenomenon. A lighted candle was let down, and the flame became agitated and thrown in one direction at a depth of 30 feet, but was quite still at the bottom; where, however, it soon died out. The water is hard or limestone water.

Rocks of volcanic formation would seem to afford favourable opportunities for the formation of ice. Scrope mentions this fact in an account of the curious district called Eiffel or Eifel, in Rhenish Prussia, which was published originally in the 'Edinburgh Journal of Science,'[147] and has since been translated in Keferstein's Deutschland.[148] The village of Roth, near Andernach, is built on a current of basalt, derived from the cone above it, which has at some time sent down a stream of lava to the north and west. A small cavern near the village, forming the mouth of a deep fissure in the lava-stream, half-way up the cone, displays a phenomenon which the writer says he has often observed in volcanic formations. The floor of the cavern was covered with a crust of ice at the time of his visit, about noon on a very hot day in August. The peasants report that there is always ice in summer, and never in winter, when the sheep retreat to the cave on account of its warmth. Steininger[149] found a thickness of 3 feet of ice on September 19, 1818, but it was evidently in a melting state, and the thermometer stood at 36·5 F. in the cavern. He describes it as possessing a narrow entrance facing north, entirely sheltered from the sun by lava-rocks, and by the trees of a wood which covers the cone of scoria.

Scrope believes that this is the mouth of one of the arched galleries so frequently met with under lava in Iceland, Bourbon, and elsewhere; and on this he founds his explanation of the phenomenon. If the other extremity is connected with the external air at a much lower level, a current of air must be constantly driven up this gallery, and in its passage will be dried by the absorbent nature of the rock--which is perhaps partly owing to the sulphuric or muriatic acid it contains[150]--- and the evaporation caused by this current produces a coating of ice on the floor of the grotto, where there is a superficial rill of water. The more rarified the lower external air, the more rapid will be the current of cool air; and, therefore, the greater the evaporation. The winter phenomenon is to be explained by the fact that the current of air will be about the mean annual temperature of the district,

---

[147] v. 154.

[148] iv. 300.

[149] *Die erlöschenen Vulkane in der Eifel*, S. 59.

[150] Dr. Gmelin, of Tubingen, detected the presence of ammonia both in clinkstone lava and in columnar basalt (*American Journal of Science*, iv. 371).]

taking its temperature, in fact, from the rocks through which it passes; and, therefore, by contrast the grotto will appear warm.

The same writer mentions a similar example of summer ice in Auvergne.[151] There is a natural grotto in the basalt near Pont Gibaud, some miles to the north-west of Clermont, in which a small spring is found partly frozen during the greatest heats of summer, while the water is said to be warm in winter; probably, Scrope observes, only seeming to be warm by contrast with the external temperature. The water is apparently frozen by means of the powerful evaporation produced by a current of very dry air proceeding from some long fissures or arched galleries which communicate with the cave. In this case also the writer suggests that the air owes its dryness to the absorbent qualities of the lava through which it passes: he repeats, too, the remark that the phenomenon is of common occurrence in caverns in volcanic districts.[152]

There is a remarkable instance of ice occurring under lava, near the *Casa Inglese* on Mount Etna, which it may be as well to mention, though the causes of its existence have probably nothing in common with the phenomena of ice-caves, or summer ice. An account of it is to be found in Sir Charles Lyell's 'Elements of Geology.'[153] It appears that the summer and autumn of 1828 were so hot, that the artificial ice-houses of Catania and the adjoining parts of Sicily failed. Signer M. Gemmellaro had long believed that a small mass of perennial ice at the foot of the highest cone of Etna was only a part of a large and continuous glacier covered by a lava current, and from this he expected to derive an abundant supply of ice. He procured a large body of workmen, and quarried into the ice; but though he thus proved the superposition of lava for several hundred yards, the ice was so hard, and the expense of quarrying consequently so great, that the works were abandoned. This was on the south-east of the cone, not far from the *Casa Inglese*. Sir Charles Lyell suggests that, probably, at the commencement of some eruption, a large mass of snow has been thickly

---

[151] Geology and Extinct Volcanoes of Central France, p. 60 (second edition).

[152] Mr. William Longman has informed me that some years ago he had ice given him in summer, when he was on a visit to the inspector of mines at Pont Gibaud, and he was told that it was formed in a neighbouring cavern during the hot season.

[153] Original edition of 1830, i. 369.

covered with volcanic sand, showered upon it before the arrival of the lava itself. This sand is a non-conductor of heat, and would therefore tend to preserve the snow from complete fusion when the hot lava-stream passed over it, and thus the existence of the underground glacier may be explained. The peasants of the district are so well acquainted with the non-conducting properties of volcanic sand, that they secure an annual store of snow, for providing water in summer, by strewing a layer of sand a few inches thick upon a field of snow, thus effectually shutting out the heat of the sun. It is curious that when De Saussure visited Chamouni for the first time, his attention was arrested by the sight of women sowing what seemed to be grain of some kind in the snow; but, on enquiring, he found that it was only black earth, which the inhabitants spread on the snow in spring, in order to make it disappear sooner. He was told that snow thus treated would melt a fortnight or three weeks before the ordinary time for its disappearance in the valley; but it will be seen that this does not contradict the theory of the Sicilian peasants.[154]

Sir Charles Lyell adds that, after what he saw on Mount Etna, he should not be surprised to find layers of glacier and lava alternating in some parts of Iceland.

Something similar was observed by Von Kotzebue, near the sound which bears his name.[155] His party was encamped on a large plain covered with moss and grass, when they discovered a fissure which revealed the fact that the moss and grass were but a thin coating on a layer of ice a hundred feet thick. This was not mere frozen ground, but aboriginal ice; for, in the ice which formed the walls of the fissure, they found the bones and teeth of mammoths embedded.

The frozen soil of Jakutsk, in Siberia, has for many years attracted considerable attention. The ordinary law of increase of temperature in descending below the surface of the earth would appear, however, to be only modified here; for it is found in sinking a well which has afforded

---

[154] See Professor Tyndall's *Glaciers of the Alps*, for an account of glacier-tables, sand-cones, &c. Anyone who has walked on a glacier will have noticed the little pits which any small black substance, whether a stone or a dead insect, sinks for itself in the ice.

[155] Gilbert, *Annalen*, lxix. 143.

opportunities for observing the state of the soil, that the temperature gradually increases with the depth.[156]

Two ice-caverns were examined by Georgi, in the course of his travels in Russia.[157] One occurs near the mines of Lurgikan, on the east side of a hill about 450 feet high, not far from the confluence of the Lurgikan stream with the Schilka (a tributary of the Amur), in the province of Nertschinsk. In the course of driving an adit in one of the lead-mines, in the year 1770, the workmen were struck by the hollow sound given forth by the rock, and, on investigation, they found an immense grotto or fissure, of which the entrance was so much blocked up by ice that they had much difficulty in sliding down by means of ropes. The fissure extended under the hill, in a direction from north to south, and was 130 fathoms long, from 1 to 8 broad, and from 3 to 12 high. Where it approached nearest the surface, the thickness of the roof was about 10 fathoms. The rock is described by Georgi as *quarzig, bräunlich, und von einem starken Kalkschuss*. He found the greater part of the walls covered with ice, and many pillars and pyramids of ice on the floor. The cold was moderate, and was said to be much the same in summer and winter. Patrin has given a fuller description of the same cavern in the *Journalde Physique*.[158] The lead-mine is in limestone rock, containing a third part of clay. The entrance to the glacière was still difficult at the time of his visit, and it was necessary to use a rope, and also to cut steps, for the descent was made along a ridge of ice with almost perpendicular sides. The spectacle presented by the decoration of

[156] According to the latest accounts I have been able to obtain, a temperature of 29·75° F. had already been reached some years ago; the temperature, a few feet from the surface, being 14° below freezing. The soil here only thaws to a depth of 3 feet in the hottest summer. Sir R. Murchison wrote to Russia, in February last, for further information regarding this well. Since I wrote this, Sir Roderick Murchison has applied to the Secretary of the Imperial Academy of St. Petersburg for further information respecting the investigations at Jakutsk. The Secretary gives a reference to Middendorff's *Sibirische Reise*, Bd. iv. Th. i., 3te Lieferung, *Klima*, 1861. I have only been able to find the edition of 1848-51; but in that edition, under the heading *Meteorologische Beobachtungen*, elaborate tables of the meteorological condition of Jakutsk are given (i. 28-49). Also, under the heading *Geothermische Beobachtungen*, very careful information respecting the frozen earth will be found (i. 157, &c., and 178, &c.). The point at which a temperature of 32° will be attained, is reckoned variously at from 600 to 1,000 feet below the surface.

[157] Reise im Russischen Reich, i. 359; St. Petersburg, 1772.

[158] xxxviii. 231 (an. 1791), in an article called *Notice minéral, de la Daourie.*

the roof was remarkably beautiful, long festoons and tufts of ice hanging down, light and brilliant as silver gauze: this ice was supposed to be formed from the abundant vapours of the beginning of winter, and resembled glass blown to the utmost tenuity. It was crystallised, too, in a wonderful manner. Patrin found long bundles of hexahedral tubes, the walls of which were formed of transverse needles: the diameter of these tubes was from two to six lines only, but at the lower extremities they opened out into hollow six-sided pyramids, more than an inch in diameter, so that the festoons, sometimes as large round as a man, presented terminal tufts of some feet in diameter, which glittered like diamonds under the influence of the torches. Towards the farther end of the fissure, stalactites of solid ice were found, displaying all the forms and more than all the beauty of limestone stalactites. The other instance mentioned by Georgi occurred in the mines of Serentvi, where two of the levels yielded perennial ice, and were thence (Georgi says) called *Ledenoi*. A spring of water flowed from the rock at a depth of thirty fathoms below the surface, and was promptly frozen into a coating of ice a foot thick. Patrin[159] visited Serentvi, but he did not observe any ice in the mines. He believed the rock to be very ancient lava.

Reich[160] mentions a cavern on Mount Sorano which contains ice, quoting Kircher;[161] but he seems to have misinterpreted his author's Latin.[162] He also refers to the existence of ice in the mines of Herrengrund in Hungary, and Dannemora in Sweden. Kircher, who has the credit of having been the first to call attention to the increase of temperature in the earth, made full enquiries into the temperature of the mines at Herrengrund, but he was not informed of the existence of ice.[163] Townson

---

[159] L.c., p. 236.

[160] *Beobachtungen*, &c., 194.

[161] *Mundus Subterraneus*, i. 220 (i. 239, in the edition of 1678).

[162] 'Vidi ego in Monte Sorano cryptam veluti glacie incrustatam, ingentibus in fornice hinc inde stiriis dependentibus, e quibus vicini mentis accolæ pocula æstivo tempore conficiunt, aquæ vinoque quæ iis infunduntur refrigerandis aptissima, extremo rigore in summas bibentium delicias commutato.'

[163] Both here and at Schemnitz, Kircher made particular enquiries on a subject of which scientific men have altogether lost sight. At Schemnitz he asked the superintendent, *an comparcant Dæmunculi vel pygmæi in fodinis?—respondit affirmative, et narrat plura exempla*; and at

visited these mines in the course of his travels in Hungary, and neither does he make any mention of ice in connection with them. He describes them as lying south of Teplitz, in a limestone district, with sandstone in the more immediate neighbourhood. The mines themselves (copper mines) are in a kind of mica-schist, which the people call granite. The superintendent of mines informed Reich that one of the shafts is called the ice-mine, from the fact that when the workmen attempted to drive a gallery from south to north, they came upon ice filling up the interstices of the *Haldenstein*, within five fathoms of the commencement of the gallery. The temperature was so low, and the expense caused by the frozen mass so great, that the working was stopped.

The iron mines of Dannemora, eleven leagues from Upsal, contain a large quantity of ice, according to a manuscript account by Mr. Over-assessor-of-the-board-of-mines Winkler:[164] Jars, however, in his *Voyages Métallurgiques*,[165] gives a full description of them without mentioning the existence of ice. He states that ice is found in the mines of Nordmarck, three leagues from Philipstadt in Wermeland, a province of Sweden: these mines are merely numerous shafts sunk in the earth, reaching to the bottom of the vein of ore, so that they are fully exposed to the light, and yet the walls of the shafts become covered with ice at the end of winter, which remains there till the middle of September. Jars believed that, if it were not for the heat caused by blasting, and by the presence of the workmen, the ice would be perennial. Humboldt[166] speaks of the ice in these mines and on the Sauberg. Reich states that ice is found in the mill-stone quarry of Nieder-Mendig, quoting Karsten's *Archiv für Bergbau*.[167] The ice is found in the hottest days of summer, although the interior of the quarry is connected with the outer air by many side shafts. The porous nature of the stone is assigned as the cause of the phenomenon. Daubeny (On

---

Herrengrund, *utrum appareant Dæmunculi seu pygmæi?—respondit tales visos fuisse, et auditos pluries*. (Edition of 1678, ii. 203, 205.)

[164] Reich, 199.

[165] i. 108 (Lyon, 1794).

[166] *Ueber die unterirdischen Gasarten*, 101.

[167] xvii. 386.

Volcanoes) describes the remarkable basaltic deposits at Niedermennig--as he spells it--but says nothing of the existence of ice.

Daubuisson[168] speaks of a *Schneegrube*, on a summit of the *Riesengebirge*, in Silesia, 4,000 feet above the sea; but such holes are common enough at that elevation, and I have seen two or three remarkable instances on the Jura, within the compass of one day's walk. Voigt[169] describes an *Eisgrube* in the Rhöngebirge, on the *Ringmauer*, the highest point of the *Tagstein*, where abundant ice is found in summer under irregular masses of columnar basalt. Reich had received from a forest-inspector an account of an ice-hole in this neighbourhood, called *Umpfen*, which is apparently not the same as that mentioned by Voigt.

In the Saxon Erzgebirge there are three points remarkable for their low temperature,[170] in addition to the mines on the Sauberg mentioned above. These are the *Heinrichssohle*, in the Stockwerk at Altenberg, where the mean of two years' observations gives the temperature 0°·54 F. lower at a depth of 400 feet than at the surface; the adit of *Henneberg*, on the Ingelbach, near Johanngeorgenstadt, where the temperature was again 0°·54 F. lower than in shafts some hundred feet higher; and the *Weiss Adler* adit, on the left declivity of the valley of the Schwarzwasser, above the Antonshütte. It would appear that there are local causes which affect the temperature in the Erzgebirge, for Reich found that in several places the mean temperature of the soil was higher than that of the air: for instance—

	Soil.	Air	Height above the sea.
Altenberg ...	42·732° Fahr.	41·27°	2,450 feet
Markus Röhling ...	43·542° "	41·832°	1,870 "
Johanngeorgenstadt. ...	43·115° "	41·09°	2,460 "

The temperature at Markus Röhling is peculiarly anomalous, considering the elevation of the surface above the sea.

---

168 Mém. sur les Basaltes de la Saxe, p. 147.

169 Mineralog. Reisen, ii. 123.

170 Reich, 200, 201; Bischof, Physical Researches on the Internal Heat of the Globe, 46, 47.

There is said to be an ice-cave in Nassau, but I have been unable to obtain any account of it, unless it be the same as the *ice-field* mentioned on page 303.

There is a cave in the south-east of Hungary[171] which presents the same features as several of the glacières I have visited. It is called the Ice-hole of Scherisciora, and is described as lying in the Jura-kalk, at a distance of 2-1/2 hours north-east from the forest-house of Distidiul. The approach is by ladders, down a pit 30 fathoms wide and 24 deep; and when the bottom of this pit is reached, an entrance is found to the cave in the north wall, in the neighbourhood of which is congealed snow which shortly becomes ice. The floor of the first chamber is composed of glacier-ice, separated from the side walls by a cleft from 1 to 3 feet wide, where it shows a depth of from 4 to 6 feet; it is as smooth as glass, and about 6 fathoms from the entrance a cone of ice stands upon it, 8 or 9 feet high. Both the floor and the cone are at once seen to be transformed remains of ancient masses of snow, and are of a dirty yellow colour.

At the back of this chamber, a narrow passage opens towards the interior of the mountain, and winds steeply down with a height of 4 feet, and a length of a few fathoms, till a magnificent dome is reached, on the beauties of which Herr Peters becomes eloquent. The floor is so smooth that crimpons are necessary, and stalagmites and stalactites of ice are found in rich profusion, the latter being generally formed on small limestone stalactites, while the former have no such nucleus.

There is another opening near the original entrance to the cave, a sort of fissure covered with elegant forms of ice, leading to a steep shaft. The imperial forester of Topfanalva was bold enough to let himself down the slope of ice which formed the edge of the shaft, on a rope ladder 60 feet long, notwithstanding the difficulty of grasping the iron steps which of course lay pressed on to the ice; but when he had descended about 30 feet, the shaft became perpendicular, and stones thrown in showed a very

---

[171] Peters, *Geologische und mineralogische Studien aus dem sudöstlichen Ungarn*, in the *Sitzungsberichte der kais. Ak. in Wien*, B. xliii., 1te Abth., S. 435. See also pages 394 and 418 of the same volume (year 1861).]

considerable depth. There appeared to be no sound of water in the abyss below.

Both entrances, that to the shaft as well as that to the second chamber, were ornamented with delicate ice crystals, which occurred both on the limestone stalactites and on the walls, and presented almost the appearance of plants of cauliflower. The ice-floor of the first chamber is described as consisting of a 'coarse-grained' material.

In the south-east of Servia, on the western slope of Mount Rtagn, is a pit 20 feet in diameter, and 40 or 50 feet deep, the bottom of which is reached by a succession of trunks of trees with the branches lopped off, a sort of ladder called *stouba* by the natives.[172] The peasants assert that the snow and ice disappear from this pit in September, and do not reappear before June. The Swiss peasants have never yet got so far as to say that the *snow* in their pits disappears in winter and returns in summer. Boué[173] found the temperature of the bottom of the pit to be 28°·4 F., while that of the air outside was 76° F. The same writer[174] mentions a source in a mill-stone quarry in Bosnia which is frozen till the end of June.

[172] Such ladders are in ordinary use in the Jura.

[173] *Turquie d'Europe,* i. 132 (he quotes himself as i. 180, in the *Sitzungsb, der k. Ak. in Wien*, xlix. l.324).

[174] L. c., p, 521.

Chapter 17

HISTORY OF THEORIES RESPECTING THE CAUSES OF SUBTERRANEAN ICE

The only glacière which is in any sense historical, is that near Besançon; and a brief account of the different theories which have been advanced in explanation of the phenomena presented by it, will include almost all that has been written on ice-caves.

The first mention I have found of this cave is contained in an old history of the Franche Comté of Burgundy, published at Dôle in 1592, to which reference has been already made. Gollut, the author, speaks more than once of a *glacière* in his topographical descriptions, and in a short account of it he states that it lay near the village of *Leugné*, which I find marked in the Delphinal Atlas very near the site of the Chartreuse of Grâce-Dieu; so that there can be no doubt that his glacière was the same with that which now exists. His theory was, that the dense covering of trees and shrubs protected the soil and the surface-water from the rays of the sun, and so the cold which was stored up in the cave was enabled to withstand the attacks of the heat of summer.[175] In the case of many of the

[175] As Gollut's phraseology is peculiar, it may be as well to reproduce his account of the cave:— 'Je ne veux pas omettre toutefois (puisque je suis en ces eaux) de mettre en memoire la commodité que nature hat doné à quelques delicats, puis qu'au fond d'un mõntagne de Leugné, la glace (*glasse* in the index), se treuve en esté, pour le plaisir de ceux qui aim[~e]t

glacières, there can be no doubt that this idea of winter cold being so preserved, by natural means, as to resist the encroachments of the hotter seasons, is the true explanation of the phenomenon of underground ice.

The next account of this glacière is found in the History of the Royal Academy of Sciences (French), under the year 1686,[176] but no theory is there suggested. The writer of the account states that in his time the floor of the cave was covered with ice, and that ice hung from the roof in festoons. In winter the cave was full of thick vapours, and a stream of water ran through it. The ice had for long been less abundant than in former times, in consequence of the felling of some trees which had stood near the entrance.

The Academy received in the same year another letter on this subject, confirming the previous account, and adding some particulars. From this it would seem that people flocked from all sides to the glacière with waggons and mules, and conveyed the ice through the various parts of Burgundy, and to the camp of the Saone; not thereby diminishing the amount of ice, for one hot day produced as much as they could carry away in eight days. The ice seemed to be formed from a stream which ran through the cave and was frozen in the summer only. The writer of this second account saw vapours in the glacière (the editor of the *Histoire de l'Académie* does not say at what season the visit to the cave took place), and was informed that this was an infallible sign of approaching rain; so much so, that the peasants were in the habit of determining the coming weather by the state of the grotto.

In 1712, M. Billerez, Professor of Anatomy and Botany in the University of Besançon, communicated to the Academy[177] an account of a visit made by him to this cave in September 1711. He found 3 feet of ice

---

a boire frais. Néanmoins dans ce t[~e]ps cela se perd, nõ pour autre raison (ainsi que íe pense) que pour ce que lon hat dépouillé le dessus de la mõtagne d'une époisse et aulte fustaie de bois, qui ne permettoit pas que les raions du soleil vinsent échauffer la terre et déseicher les distillations, que se couloi[~e]t iusques au plus bas et plus froid de la montagne: ou (par l'antipéristase) le froid s'epoississoit, et se reserroit, contre les chaleurs, entornantes et environnantes le long de l'esté, toute la circonference extérieure du mont.'—*Histoire*, &c. p. 87.

[176] *Hist. de l'Acad.*, t. ii., p. 2.

[177] *Hist. de l'Acad.*, an 1712, p. 20.

on the floor of the cave, in a state of incipient thaw, and three pyramids, from 15 to 20 feet high and 5 or 6 feet in diameter, which had been already considerably reduced in size by thaw. A vapour was beginning to pass out from the cave, at the highest part of the arch of entrance; a phenomenon which, he was told, continued through the winter, and announced or accompanied the departure of the ice: nevertheless, the cold was so great that he could not remain in the glacière more than half an hour with any sort of comfort. The thermometer stood at 60° outside the cave, and fell to 10°[178] when placed inside; but thermometrical observations of that date were so vague as to be useless for present purposes. The ice appeared to be harder than the ordinary ice of rivers, less full of air-bubbles, and more difficult to melt.

M. Billerez enunciated a new theory to account for the phenomena presented by the cave. He observed that the earth in the immediate neighbourhood, and especially above the roof of the grotto, was full of a nitrous or ammoniac salt, and he accordingly suggested that this salt was disturbed by the heat of summer and mingled itself with the water which penetrated by means of fissures to the grotto, and so the cave was affected in the same way as the smaller vessel in the ordinary preparation of artificial ice. He had heard that some rivers in China freeze in summer from the same cause.[179]

In 1726, a further communication was made to the Academy by M. des Boz,[180] Royal Engineer, describing four visits which he had made to the grotto near Besançon at four different seasons of the year, viz., in May and November 1725, and in March and August 1726. In all cases he found the air in the cave colder than the external air,[181] and its variations in temperature corresponded with the external variations, the cold being greater in winter than in summer.

---

[178] *C'est à dire*—M. Billerez explains—*à 10 degrés au-dessous du très-grand froid.* What the 60° may be worth, I cannot say.

[179] Tournefort (*Voyage du Levant*, iii. 17) believed that the ammoniac salt, of which the earth was full in some districts near Erzeroum, had something to do with the persistence of snow on the ground there.

[180] *Hist, de l'Acad.*, an 1726, p. 16.

[181] But see on this point the experience of M. Thury, in the Glacière of S. Georges (Appendix).

M. des Boz ascribed the existence of ice in the cave to natural causes. The opening being towards the north-east, and corresponding with a gorge in the hills opposite, running in the same direction, none but cold winds could reach the mouth of the grotto. Moreover, the soil above was so thickly covered with trees and brushwood, that the rays of the sun could not reach the earth, much less the rock below. Credible persons asserted that since some of the trees had been felled, there had not been so much ice in the cave.

In order to test the presence of salt, M. des Boz melted some of the ice, and evaporated the resulting water, but found no taste of salt in the matter which remained.[182] He denied the existence of the spring of water which previous accounts had mentioned, and believed that the water which formed the ice came solely from melted snow, and from the fissures of the rock.

In 1727, the Duc de Lévi caused the whole of the ice to be removed from the cave, for the use of the army of the Saone, which he commanded. In 1743 the ice had formed again, and the grotto was subjected to a very careful investigation by M. de Cossigny, chief engineer of Besançon, in the months of August and October.[183] The thermometer he used had been presented to him by the Academy, and was very probably constructed by M. de Réaumur himself, for de Cossigny's account was sent through M. de Réaumur to the Academy, but still the observations made with it cannot be considered very trustworthy. On the 8th of August, at 7.30 A.M., the temperature in the cave was 1/2° above the zero point of this thermometer, and at 11.30 A.M. it had risen to 1° above zero. On the 17th of October, at 7 A.M., the thermometer stood at 1/2°, and at 4 P.M. it gave the same register.

M. de Cossigny found that the entrance to the cave was rather more than 150 feet above the Abbey of Grâce-Dieu, and about half a league distant by the ordinary path. A great part of his account is occupied by

---

[182] Sir Roderick Murchison's suggestion of the possible influence of salt in producing the phenomena of his ice-cave in Russia, did not, of course, proceed upon the supposition of salt actually mingling with water, but only of its increasing the evaporation of the air which came in contact with it.]

[183] Mém. présentés à l'Académie par divers Sçavans, i, 195.

contradictions of previous accounts, especially in the matter of dimensions,[184] The people of Besançon had urged him to stay only a short time in the cave, because of the sulphureous and nitrous exhalations, but he detected no symptoms of anything of that kind. The most curious thing which he saw was the soft earth which lay, and still lies, at the bottom of the long slope of ice by which the descent is made; and he subjected this to various chemical tests and processes, but could not find that it contained anything different from ordinary earth.[185]

When M. de Cossigny visited the cave, there were thirteen or fourteen columns of ice, from 6 to 8 feet high, and he was in consequence inclined to doubt the accuracy of the statement of M. Billerez, that in his time (1711) there were three columns only, from 15 to 20 feet high. But my own observation of the shape of the columns suggested that the largest of all was probably an amalgamation of several others; so that it is not unreasonable to suppose that after the Duc de Lévi removed the large columns seen by M. Billerez, a number of smaller columns were formed on the old site, and that these had not become large enough to amalgamate in 1743.

Not satisfied with these visits of August and October, M. de Cossigny visited the cave in April 1745. He found the temperature at 5 A.M. to be exactly at the freezing point, and at noon it had risen 1°. From this he concluded that the stories of the greater cold in the cave during the summer, as compared with the winter, were false.

In 1769, M. Prévost, of Geneva, visited the cave, as a young man; and in 1789, he wrote an account of his visit in the *Journal de Genève* (March), which was afterwards inserted as an additional chapter in his book on Heat.[186] He believed that one or two hundred *toises* was the utmost that could be allowed for the height of the hill in which the glacière lies,--a sufficiently vague approximation. He rejected the idea of salt as the cause

---

[184] A long account was published in a history of Burgundy, printed at Dijon, in quarto, in 1737, which I have not been able to find. It was from the same source as the account in the Hist. of the Academy, in 1726.

[185] I took this earth to be a collection of the particles carried down the slope of ice by the heavy rains of the month preceding my visit. M. de Cossigny speaks of the abundant rains of July, his visit being in August.

[186] *Recherches sur la Chaleur*; Geneva and Paris, 1792.

of ice, and came to the conclusion that the cave was in fact nothing more than a good natural ice-house, being protected by dense trees, and a thick roof of rock, while its opening towards the north sheltered it from all warm winds. He accounted for the original presence of ice as follows:--In the winter, stalactites form at the edges of various fissures in the roof, and snow is drifted on to the floor of the cave by the north winds down the entrance-slope. When the warmer weather comes, the stalactites fall by their own weight, and, lying in the drifted and congealed snow, form nuclei round which the snow is still further congealed, and the water which results from the partial thaw of portions of the snow is also converted into ice. Thus, a larger collection of ice forms in winter than the heat of summer can destroy; and if none of it were removed, it might, in the course of years, almost fill the cave. At the time of his visit (August), M. Prévost found only one column, from 6 to 8 feet high.

In 1783 (August 6), M. Girod-Chantrans visited the Glacière of Chaux (so called from a village near the glacière, on the opposite side from the Abbey of Grâce-Dieu), and his account of the visit appeared in the *Journal des Mines*[187] of Prairial, an iv., by which time the writer had become the Citizen Girod-Chantrans. He found a mass of stalactites of ice hanging from the roof, as if seeking to join themselves with corresponding stalagmites on the floor of the cave; the latter, five in number, being not more than 3 or 4 feet high, and standing on a thick sheet of ice. There was a sensible interval between this basement of ice and the rock and stones on which it reposed: it was, moreover, full of holes containing water, and the lower parts of the cave were unapproachable by reason of the large quantity of water which lay there. The thermometer stood at 35°·9 F. two feet above the floor, and at 78° F. in the shade outside. M. Girod-Chantrans determined, from all he saw and heard, that the summer freezing and winter thaw were fables, and he believed that the cave was only an instance of Nature's providing the same sort of receptacle for ice as men provide in artificial ice-houses. He was fortunate enough to obtain by chance the notes of a neighbouring physician, who had made careful observations and

[187] P. 65. Now called *Annales des Mines*.

experiments in the glacière at various seasons of the year, and a *précis* of these notes forms the most valuable part of his account.

Dr. Oudot, the physician in question, found ten columns in January 1778, the largest of which was 5-1/2 feet high. The flooring of ice was nowhere more than 15 inches thick, and the parts of the rock which were not covered with ice were perfectly dry. The thermometer--M. Girod-Chantrans used Réaumur, so I suppose that he gives Dr. Oudot's observations in degrees of Réaumur, though some of the results of that supposition appear to be anomalous--gave 22° F. within the cave, and 21° F. outside.

In April of the same year, the large column had increased in height to the extent of 13 inches; and the floor of ice on which it stood was 1-1/2 inch thicker, and extended over a larger area than before; the thermometer stood at 36°·5 F. and 52° F. respectively in the same positions as in the former case. In July, the large column had lost 6 inches of its height, and the thermometer gave 38°·75 F. and 74°·75 F.

In October, the large column was only 3 feet high, and many of the others had disappeared, while their pedestal had become much thinner than it had been in the preceding months. There was also a considerable amount of mud in the cave, brought down apparently by the heavy rains of autumn. The thermometer gave 37°·6 F. and 63°·5 F.

On the 8th of January, 1779, there were nine columns of very beautiful ice, and one of these, as before, was larger than the rest, being 5 feet high and 10 feet in circumference. The temperatures were 21° F. and 16°·15 F. in the cave and in the open air respectively.

Tradition related that, before the removal of the ice in 1727, one of the columns reached the roof, (Prévost calculated the limits of the height of the cave at 90 and 60 feet,) and this suggested to Dr. Oudot the idea of placing stakes of wood in the heads of the columns he found in the cave, in the hope that ice would thus collect in greater quantities under the fissures of the roof. Accordingly, he made holes in three of the columns, and established stakes 4, 5, and 10 feet high, returning on the 22nd of February, after an interval of six weeks, to observe the result of his experiment. He found the two shorter stakes completely masked with ice, forming columns a foot in diameter; and the longest stake, though not entirely concealed by the ice which had collected upon it, was crowned with a beautiful capital of perfectly transparent ice. The columns which had no stakes fixed upon them had also increased somewhat in size, but not nearly in the same proportion as those which were the subject of Dr. Oudot's experiment. The thermometer on this day gave 29°·5 F. and 59° F. as the temperatures.

It may be remembered that I found one very beautiful column, far higher than any of those mentioned by Dr. Oudot, and higher than those which M. Billerez saw, formed upon the trunk and branches of a fir-tree. I have now no doubt that the peculiar shape of another--the largest of the three columns which were in the cave at the time of my visit--is due to the fact of its being a collection of several smaller columns, which have in course of time flowed into one as they increased separately in bulk, and that its height has been augmented by a device similar to that adopted by Dr. Oudot. The two magnificent capitals which this column possessed, as well as the numerous smaller capitals which sprang from its sides, will thus be completely accounted for.

One more account may be mentioned, before I proceed to the theory which has found most favour in Switzerland of late years. M. Cadet published some *Conjectures* on the formation of the ice in this cavern, in the *Annales de Chimie,* Nivôse, an XI.[188] He saw the cave in the end of September 1791, and found very little ice--not a third of what there had been a month before, according to the account of his guide. The *limonadier* of a public garden in Besançon informed him that the people of that town resorted to the glacière for ice when the supplies of the artificial ice-houses failed, and that they chose a hot day for this purpose, because on such days there was more ice in the cave. Ten *chars* would have been sufficient to remove all the ice M. Cadet found, and the air inside the cave seemed to be not colder than the external air; but, nevertheless, M. Cadet believed the old story of the greater abundance of ice in summer than in winter, and he attempted to account for the phenomenon.

---

[188] T. xlv. p. 160.

The ground above and near the cave is covered with beech and chestnut trees, and thus is protected from the rays of the sun. The leaves of these trees give forth abundant moisture, which has been pumped up from their roots; and as this moisture passes from the liquid to the gaseous state, it absorbs a large quantity of caloric. Thus, throughout the summer, the atmosphere is incessantly refrigerated by the evaporation produced by the trees round the cave; whereas in winter no such process goes on, and the cave assumes a moderate temperature, such as is usually found in ordinary caves. Unfortunately for M. Cadet's theory, the facts are not in accordance with his imaginary data, nor yet with his conclusions. He adds, on the authority of one of his friends, that the intendant of the province, M. de Vanolles, wishing to preserve a larger amount of ice in the cave, built up the entrance with a wall 20 feet high, in which a small door was made, and the keys were left in the hands of the authorities of the neighbouring village, with orders that no ice should be removed. The effect of this was, that the ice diminished considerably, and they were obliged to pull down the wall again. M. Cadet saw the remains of the wall, and the story was confirmed by the Brothers of Grâce-Dieu. It would be very interesting to know at what season this wall was built, and when it was pulled down. If my ideas on the subject of ice-caves are correct, it would be absolutely fatal to shut out the heavy cold air of winter from the grotto.

In 1822, M.A. Pictet, of Geneva, took up the question of natural glacières, and read a paper before the Helvetic Society of Natural Sciences,[189] describing his visits to the caves of the Brezon and the Valley of Reposoir. In order to explain the phenomena presented by those caves, M. Pictet adopted De Saussure's theory of the principle of *caves-froides*, rendering it somewhat more precise, and extending it to meet the case of ice-caves. It is well known that, in many parts of the world, cold currents are found to blow from the interstices of rocks; and these are utilised by neighbouring proprietors, who build sheds over the fissures, and so secure a cool place for keeping meat, &c. Examples of such currents are met with near Rome (in the *Monte Testaceo*), at Lugano, Lucerne (the caves of

[189] Bibliothèque Universelle de Genève, Première Série, t. xx.

Hergiswyl), and in various other districts. It is found that the hotter the day, the stronger is the current of cold air; in winter the direction of the current is changed, and it blows into the rock instead of out from it.[190] De Saussure's theory, as developed by M. Pictet, was no doubt satisfactory, so far as it was used to account for the phenomenon of 'cold-caves,' but it seems to be insufficient as an explanation of the existence of large masses of subterranean ice; of which, by the way, De Saussure must have been entirely ignorant, for he makes no allusion to such ice, and the temperatures of the coldest of his caves were considerably above the freezing point.

Pictet represents the case of a cave with cold currents of air to be much the same as that of a mine with a vertical shaft, ending in a horizontal gallery of which one extremity is in communication with the open air, at a point much lower, of course, than the upper extremity of the shaft. The cave corresponds to the horizontal gallery, and the various fissures in the rock take the place of the vertical shaft, and communicate freely with the external air. In summer, the columns of air contained in these fissures assume nearly the temperature of the rock in which they rest, that is to say, the mean temperature of the district, and therefore they are heavier than the corresponding external columns of air which terminate at the mouth of the cave; for the atmosphere in summer is very much above the mean temperature of the soil, or of the interior of the earth at moderate depths. The consequence is, that the heavy cool air descends from the fissures, and streams out into the cave, appearing as a cold current; and the hotter the day is--that is, the lighter the columns of external air--the more violent will be the disturbance of equilibrium, and therefore the more palpable the cold current. Naturally, in this last case, the air which enters by the upper orifices of the fissures is more heated, to begin with, than on cooler days; but external heat so very slightly affects the deeper parts of the fissures, that the columns of air thus introduced are speedily impressed with the mean temperature of the district. In winter, the external columns of air are as much heavier than the columns in the fissures as they are lighter in

---

[190] See De Saussure's account of his numerous observations of such caves in the *Voyage dans les Alpes*, sections 1404-1415.

summer; and so cold currents of air blow from the cave into the fissures, though such currents are not of course colder than the external air. Thus the mean temperature of the cave is much lower than that of the rock in which it occurs; for the temperature of the currents varies from the mean temperature of the rock to the winter temperature of the external atmosphere.

The descending columns of warmer air, in summer, must to some extent raise the temperature of the fissures above that which they would otherwise possess, that is, above the mean temperature of the place; but that may be considered to be counteracted by the corresponding lowering of the temperature of the fissures by the introduction of cold air from the cave in winter. By a similar reasoning, it will be seen that for some time after the spring change of direction in the currents takes place, the temperature of the cave will be less than would have been expected from a calculation founded on the true mean temperature of the rock through which the fissures pass. This, together with the fact of the porous nature of the rock in which most of the curious caves in the world occur, which allows a considerable amount of moisture to collect on all surfaces, and thereby induces a depression of temperature by evaporation, may be held to explain the presence of a greater amount of cold than might otherwise have been fairly reckoned upon in ice-caves.

The idea of cold produced by evaporation Pictet took up warmly, believing that when promoted by rapid currents of air it would produce ice in the summer months; and he thus explained what he understood to be the phenomena of glacières. But it will have been seen, from the account of the caves I have visited, that the glacières are more or less in a state of thaw in the summer; and M. Thury's observations in the winter prove conclusively that they are then in a state of utter frost, so that the old belief with respect to the season at which the ice is formed may be supposed to have been exploded. The facts recorded by Mr. Scrope[191] would appear to depend upon the peculiar nature of rocks of volcanic formation; and I am inclined to think there is very little in common between such instances as he

[191] P. 271.

mentions and the large caves filled with ice which are to be found in the primary or secondary limestone.

One of De Saussure's experiments, in the course of his investigation of the phenomena and causes of cold currents in caves, is worth recalling. He passed a current of air through a glass tube an inch in diameter, filled with moistened stones, and by that means succeeded in reducing the temperature of the current from 18° C. to 15° C.; and when the refrigerated current was directed against a wet-bulb thermometer, it fell to 14° C., thus showing a loss of 7°·2 F. of heat. No one can see much of limestone caverns without discovering that the surfaces over which any currents there may be are constrained to pass, present an abundance of moisture to refrigerate the currents; and it is not unreasonable to suppose that the large number of evaporating surfaces, which currents passing through heaps of débris--such as the basaltic stones described on page 261--come in contact with, are the main cause of the specially low temperature observed under such circumstances.

Pictet's theory, however, did not convince all those into whose hands his paper fell, and M.J. Deluc wrote against it in the *Annales de Chimie et de Physique* of the same year, 1822.[192] Deluc had not seen any glacière, but he was enabled to decide against the cold-current theory by a perusal of Pictet's own details, and of one of the accounts of the cave near Besançon. He objected, that in many cases the ice is found to melt in summer, instead of forming then; and also, that in the Glacière of S. Georges, which Pictet had described, there was no current whatever. Further, in all the cases of cold currents investigated or mentioned by De Saussure, the presence of summer ice was never even hinted at, and the lowest temperatures observed by him were considerably above the freezing point. I may add, from my own experience, that on the only occasions on which I found a decided current in a glacière--viz., in the Glacière of Monthézy, and that of Chappet-sur-Villaz,--there was marked thaw in connection with the current. In the latter case, the channel from which the current came was filled with water; and in the former, water stood on the surface of the ice.

---

[192] xxi. 113.

The view which Deluc adopted was one which I have myself independently formed; and he would probably have written with more force if he had been acquainted with various small details relating to the position and surroundings of many of the caves. The heavy cold air of winter sinks down into the glacières, and the lighter warm air of summer cannot on ordinary principles of gravitation dislodge it, so that heat is very slowly spread in the caves; and even when some amount of heat does reach the ice, the latter melts but slowly, for ice absorbs 60° C. of heat in melting; and thus, when ice is once formed, it becomes a material guarantee for the permanence of cold in the cave.

For this explanation to hold good, it is necessary that the level at which the ice is found should be below the level of the entrance to the cave; otherwise the mere weight of the cold air would cause it to leave its prison as soon as the spring warmth arrived. In every single case that has come under my observation, this condition has been emphatically fulfilled. It is necessary, also, that the cave should be protected from direct radiation, as the gravitation of cold air has nothing to do with resistance to that powerful means of introducing heat. This condition, also, is fulfilled by nature in all the glacières I have visited, excepting that of S. Georges; and there art has replaced the protection formerly afforded by the thick trees which grew over the hole of entrance. The effect of the second hole in the roof of this glacière is to destroy all the ice which is within range of the sun. A third and very necessary condition is, that the wind should not be allowed access to the cave; for if it were, it would infallibly bring in heated air, in spite of the specific weight of the cold air stored within. It will be understood from my descriptions of such glacières as that of the Grand Anu, of Monthézy, and the Lower Glacière of the Pré de S. Livres, how completely sheltered from all winds the entrances to those caves are. There can be no doubt, too, that the large surfaces which are available for evaporation have much to do with maintaining a somewhat lower temperature than the mean temperature of the place where the cave occurs. This had been noticed so long ago as Kircher's time; for among the answers which his questions received from the miners of Herrengrund, we find it stated that, so long as mines are dry, the deeper they are the hotter; but if they have water, they are less warm, however deep. From the mines of Schemnitz he was informed that, so long as the free passage of air was not hindered, the mines remained temperate; in other cases they were very warm. Another great advantage which some glacières possess must be borne in mind, namely, the collection of snow at the bottom of the pit in which the entrance lies. This snow absorbs, in the course of melting, all heat which strikes down by radiation or is driven down by accidental turns of the wind; and the snow-water thus forced into the cave will, at any rate, not seriously injure the ice. It is worthy of notice that the two caves which possess the greatest depth of ice, so far as I have been able to fathom it, are precisely those which have the greatest deposit of snow; and the ice in a third cave, that of Monthézy, which has likewise a large amount of snow in the entrance-pit, presents the appearance of very considerable depth. The Schafloch, it is true, which contains an immense bulk of ice, has no snow; but its elevation is great, as compared with that of some of the caves, and therefore the mean temperature of the rock in which it occurs is less unfavourable to the existence of ice.

I believe that the true explanation of the curious phenomena presented by these caves in general, is to be found in Deluc's theory, fortified by such facts as those which I have now stated. The mean temperature of the rock at Besançon, where the elevation above the sea is comparatively so small, renders the temptation to suggest some chemical cause very strong.

The question of ice in summer where thaw prevails in winter, may fairly be considered to have been eliminated from the discussion of such caves as I have seen, in spite of the persistent assertions of some of the peasantry. The observations, however, in caverns in volcanic formations, and in basaltic débris, are so circumstantial that it is impossible to reject them; and in such cases a theory similar to that enunciated by Mr. Scrope[193] seems to be the only one in any way satisfactory, though I have not heard of such marvellous results being produced elsewhere by evaporation. One observer, for instance, of the cavern near the village of Both, in the Eiffel, found a thickness of 3 feet of ice; and in that case it was

[193] P. 271.

melting in summer, instead of forming. In some cases it has been suggested that the length of time required for external heat or cold to penetrate through the earth and rock which lie above the caves is sufficient to account for the phenomenon of summer frost and winter thaw. Thus, it is said, the thickness of the superincumbent bed may be such that the heat of summer only gets through to the cave at Christmas, and then produces thaw, while in like manner the greatest cold will reach the cave in mid-summer. But there is a fatal objection to this idea in the fact that the invariable stratum--i.e., the stratum beyond which the annual changes of external temperature are not felt--is reached about 60 feet below the surface in temperate latitudes,[194] while at the tropics such changes are not felt more than a foot below the surface. Humboldt calculated that in the latitude of central France the whole annual variation in temperature at a depth of 30 feet would not amount to more than one degree.[195]

---

194 Daubuisson estimated the depth in question at from 46 to 61 feet, while Kupffer put it at 77 feet.

195 De Saussure found a variation of 2°·25 F. at a depth of 29·5 feet; but this was in a well, where the influence of the atmosphere was allowed to have effect. Naturally, the fissures which there may be in the rock surrounding a cave will increase the annual variation of temperature, by affording means of easier penetration to the heat and cold.
Sir K. Murchison's cavern in Russia would seem to be entirely *sui generis*.

Chapter 18

On the Prismatic Structure of the Ice in Glacières

It was natural to suppose that the prismatic structure which I found so very general in the glacières was the result of some cause or causes coming into operation after the first formation of the ice. On this point M. Thury's visit to the Glacière of S. Georges in the spring of 1852 affords valuable information, for at that time the coating of ice on the wall, evidently newly formed, did not present the *structure aréolaire* which he had observed in his summer visit to the cave. He suggests that, since ice is less coherent at a temperature of 32° F.--which is approximately the temperature of the ice-caves during several months of the year--than when exposed to a greater degree of cold, its molecules will then become free to assume a fresh system of arrangement.[196] On the other hand, Professor Faraday has found that ice formed under a temperature some degrees below the ordinary freezing point has a well-marked crystalline structure.[197] M. Thury suggests also, as a possibility, what I have found to be the case, by frequent observations, that the prismatic ice has greater power of resisting heat than

[196] The continued extrication of latent heat by ice, as it is cooled a few degrees below 32° F., appears to indicate a molecular change subsequent to the first freezing.—*Phil. Trans.*, as quoted in the next note.

[197] See the paper 'On Liquid Diffusion as applied to Analysis,' by the Master of the Mint (*Phil. Trans.* 1861, p. 222).

ordinary ice; and on this supposition he accounts for the fact of hollow stalactites being found in the Cavern of S. Georges.[198] At the commencement of the hot season, the atmospheric temperature of the glacières rises gradually; and when it has almost reached 32° F., the prismatic change takes place in the ice, extending to a limited depth below the surface. The central parts of the stalactites retain their ordinary structure, and are after a time exposed to a general temperature rather above than below the freezing point; and thus they come to melt, the water escaping either by accidental fissures between some of the prisms, or by the extremity of the stalactite, or by some part of the surface which has chanced to escape the prismatic arrangement, and has itself melted under increased temperature.[199]

M. Héricart de Thury describes the peculiar structure of the ice which he found in the Glacière of the Foire de Fondeurle.[200] He found that the crystallised portions were very distinctly marked, displaying for the most part a six-sided arrangement; and in the interior of a hollow stalactite he found numerous needles of ice perfectly crystallised, the crystals being some triangular and some six-sided. He was unable to detect any perfect pyramid.[201] I have already quoted Olafsen's observations on the polygonal lining which he saw on the surface of the ice in the Surtshellir. The French Encyclopædia[202] relates that M. Hassenfratz saw ice served up at table at Chambéry which broke into hexagonal prisms; and when he was shown the ice-houses where it was stored, he found considerable blocks of ice containing hexahedral prisms terminated by corresponding pyramids.

---

[198] Compare the description of one of the hollow stalagmites I explored in the Schafloch, p. 145.

[199] Professor Tyndall has pointed out that, owing to the want of perfect homogeneity, some parts of a block of ice exposed to a temperature of 32° F. will melt, while others remain solid *(Phil. Trans.* 1858, p. 214). He also arrived at the conclusion (p. 219) that heat could be conducted through the substance of a mass, and melt portions of the interior, without visible prejudice to the solidity of the other parts of the mass.

[200] *Journal des Mines*, xxxiii. 157. See also an English translation of his account in the second volume of the *Edinburgh Journal of Science*.

[201] It is to be hoped that the accuracy of his scientific descriptions exceeds that of his topographical information; for he states that the glacière is two leagues from Valence, whereas it cost me six hours' drive on a level road, and five and a half hours' walking and climbing, to reach it from that town.

[202] Branch *Physique*, article *Glace*.

In vol. xv. (New Series) of the American Journal of Science,[203] an extract is given from a letter describing the 'Ice Spring' in the Rocky Mountains, which the mountaineers consider to be one of the curiosities of the great trail from the States to Oregon and California. It is situated in a low marshy 'swale' to the right of the Sweetwater river, and about forty miles from the South Pass. The ground is filled with springs; and about 18 inches below the turf lies a smooth and horizontal sheet of ice, which remains the year round, protected by the soil and grass above it. On July 12th, 1849, it was from 2 to 4 inches thick; but one of the guides stated that he had seen it a foot deep. It was perfectly clear, and disposed in hexagonal prisms, separating readily at the natural joints. The ice had a slightly saline taste,[204] the ground above it being impregnated with salt, and the water near tasting of sulphur. The upper surface of the stratum of ice was perfectly smooth.

In Poggendorff's *Annalen* (1841, Erganzsband, 517-19,--Boué, an old offender in that way, says 1842) there is an account of ice being found in the Westerwald, near the village of Frickhofen at the foot of the *Dornburg*, among basaltic débris about 500 feet above the sea.[205] Commencing at a depth of 2 feet below the surface, the ice reaches from 20 to 22 feet farther down, where the loose stones give place to dry sand. The ice is in thin layers on the stones, and is deposited in the form of clear and regular hexagonal crystals. The lateral extent through which this phenomenon obtains is from 40 to 50 feet each way, and is greater in winter than in summer. As in other cases that have been noticed in basaltic débris, the snow which falls upon the surface here is speedily melted. The *Allgemeine Zeitung* (1840, No. 309), from which the account in Poggendorff is taken, suggested that the melted snow-water which would thus run down among

---

[203] P. 146 (an. 1853).

[204] Dr. Lister experimented on sea-water in December 1684 (*Ph. Trans*, xiv. 836), and found that though it took two nights to freeze, it was much harder when once frozen than common ice, lasting for three-quarters of an hour under a heat which melted 100 times its bulk of common ice at once. It was marked with oblong squares, and had a salt taste. Ice formed from water with an admixture of sulphuric acid is said to assume a crystalline appearance.

[205] See also a pamphlet entitled *Das unterirdische Eisfeld bei der Dornburg am Südlichen Fusse des Westerwaldes*, by Thomä of Wiesbaden (32 pages, with a map of the district), published in 1841.

the interstices would readily freeze below the surface, while the heavy cold air of winter would be stored up at the lower levels, and the poor conducting powers of basaltic rock[206] would favour its permanence through the summer. The temperature of the cold current which was perceptible in the parts of the mass of débris where the ice existed was 1° R. (34°·25 F.). Nothing but a few lichens grow on the surface of the débris.

These are, I think, all the references I have met with to the prismatic structure of subterranean ice. But there is an interesting account in Poggendorff 's *Annalen*,[207] by a private teacher in Jena, of the crystalline appearance of ice under slow thaw near that town. In the winter of 1840, the Saale was frozen, and the ice remained unbroken till the middle of January, when the thermometer rose suddenly, and the river in consequence overflowed the lower grounds, and carried large masses of ice on to the fields, where it was left when the water subsided. On the 20th of January the thermometer fell again, and remained below the freezing point till the 12th of February: some of the ice did not disappear till the following month.

When the ice had lain a short time, cracks appeared on the surface exposed to the sun, and spread like a network from the edges towards the centre of the surface. At first there was no regularity in the connection of these lines, and the several meshes were of very different sizes. After a time, the larger meshes split up into smaller, and the system of network was found to penetrate below the surface, the cracks deepening into furrows, which descended perpendicularly from the surface, and divided the ice into long thin rhomboidal pillars. The surface-end of some of these pillars was strongly marked with right lines parallel to one of the sides of the mesh, and it was found that there was a tendency in the ice to split down planes through these lines and parallel to the corresponding side-plane. Parallel to the original surface of the mass of ice, the pillars broke off evenly. The side-planes had a rounded, wrinkled appearance; and their mutual inclinations--as far as could be determined--were from 105° to 115°, and from 66° to 75°. When these ice-pillars were examined by means

---

[206] But see page 262.
[207] lv. (an 1842), 472.

of polarised light, they were found to possess a feeble double-refracting power.

The writer of the article in Poggendorff suggests a question which he was not sure how to answer:--Is this appearance in correspondence with the original formation of the ice, or does it only appear under slow thaw?

It is worthy of remark, that from the 1st to the 11th of February the thermometer was never higher than 22°·8 F., and during that time fell as low as 21° below zero, i.e. 43° below the freezing point.

Professor Tyndall has informed me that in the winters of 1849, 1850, 1851, he found the banks of a river in Germany loaded with massive layers of drift-ice, in a state of thaw, and was struck by the fact that every layer displayed the prismatic structure described above, the axes of the prisms being at right angles to the surfaces of freezing. It may be, he adds, that this structure is in the first place determined by the act of freezing, but it does not develop itself until the ice thaws.

M. Hassenfratz observed an appearance in ice on the Danube at Vienna[208] corresponding to that described at Jena. He gives no information as to the state of the weather or the temperature at the time, nor any of the circumstances under which the ice came under his notice. One of the masses of ice which he describes was crystallised in prisms of various numbers of sides: of these prisms the greater part were hexahedral and irregular. Another mass was composed of prisms in the form of truncated pyramids; and in another he found quadrilateral and octahedral prisms, the former splitting parallel to the faces, and also truncated pyramids with five and six sides. He adds, that he had frequently seen in the upper valleys tufts of ice growing, as it were, out of the ground, and striated externally, but had never succeeded in discovering any internal organisation, until one evening in a time of thaw, when he found by means of a microscope that the striated tufts of ice had assumed the same structure on a small scale as that which he had observed on the Danube.

A Frenchman who was present in the room in which the Chemical Section of the British Association met at Bath, and heard a paper which I

---

[208] *Journal de Physique*, xxvi. (an 1785), 34.

read there on this prismatic structure, suggested that it was probably something akin to the rhomboidal form assumed by dried mud; and I have since been struck by the great resemblance to it, as far as the surface goes, which the pits of mud left by the coprolite-workers near Cambridge offer, of course on a very large scale. This led me to suppose that the intense dryness which would naturally be the result of the action of some weeks or months of great cold upon subterranean ice might be one of the causes of its assuming this form, and the observations at Jena would rather confirm than contradict this view: competent authorities, however, seem inclined to believe that warmth, and not cold, is the producing cause.[209]

Professor Tyndall found, in the course of his experiments on the discs and flowers produced in the interior of a mass of ice by sending a warm ray through the mass, that the pieces of ice were in some cases traversed by hazy surfaces of discontinuity, which divided the apparently continuous mass into irregular prismatic segments. The intersections of the bounding surfaces of these segments with the surface of the slab of ice formed a very irregular network of lines.[210] I am inclined, however, to think that the irregularity in these cases proved to be so much greater than that observed in the glacières, that this interior prismatic subdivision must be referred to some different cause.

---

[209] In looking through some early volumes of the *Philosophical Transactions*, I found an 'Extract of a letter written by Mr. Muraltus of Zurich (September 1668), concerning the Icy and Chrystallin Mountains of Helvetia, called the Gletscher, English'd out of Latin' (*Phil. Trans.* iv. 982), which at first looked something like an assertion of the prismatic structure of ice on a large scale. The English version is as follows:—'The snow melted by the heat of the summer, other snow being faln within a little while after, and hardened into ice, which by little and little in a long tract of time depurating itself turns into a stone, not yielding in hardness and clearness to chrystall. Such stones closely joyned and compacted together compose a whole mountain, and that a very firm one; though in summer-time the country-people have observed it to burst asunder with great cracking, thunder-like.'

[210] See the woodcut illustrating Professor Tyndall's remarks in the 148th volume of the *Philosophical Transactions* (1858, p. 214).

Chapter 19

On the Mean Temperature of the Regions in Which the Glacières Occur

Many interesting experiments have for long been carried on with a view to determine the mean temperature at various depths below the surface of the earth. The construction of Artesian wells has afforded useful opportunities for increasing the amount of our knowledge on this subject; and the well at Pregny, near Geneva,[211] and the Monk Wearmouth coal-mines, as observed by Professor Phillips while a fresh shaft was being sunk,[212] have supplied most valuable facts. Without entering into any detail, which would be an unnecessary trouble, it may be stated generally, that, under ordinary circumstances, 1° F. of temperature is gained for every 50 or 60 feet of vertical descent into the interior of the earth. I have only met with one account of an experiment made in a horizontal direction, and it is curious that the law of the increase of temperature then observed seemed to be very much the same as that determined by the mean of the vertical observations. Boussingault[213] found several horizontal adits in a

[211] Bischof, Physical Researches, 189.
[212] Philosophical Magazine, v. 446 (1834).
[213] Annules de Chimie et de Physique, liii. 2-10. See also Bischof, 136.

precipitous face of porphyritic syenite among the mountains of Marmato. In one of these adits--a gallery called Cruzada, at an elevation of 1,460 mètres--he found an increase of 1° C. of mean temperature for every 33 mètres of horizontal penetration, or, approximately, 1° F. for 60 feet.[214]

Again, observations have been made, in various latitudes, of the decrease of temperature consequent upon gradual rising from the general surface of the earth; as, for instance, in the ascent of mountains. Speaking without any very great precision, but with sufficient accuracy for ordinary purposes, 1° F. is lost with every 300 feet of ascent.[215] It is evident that this decrease will be less rapid where the slope of ascent is gradual, from such considerations as the angle at which the sun's rays strike the slope, and the larger amount of surface which is in contact with a stratum of atmosphere of any given thickness.

With these data, it is easy to arrive at some idea of the probable mean temperature of the rock containing several of the glacières I have described. The elevation of some of them has not been determined with sufficient accuracy to make the results of any calculation trustworthy; but four cases may be taken where the elevation is known--namely, the Glacières of S. Georges, S. Livres, Monthézy, and the Schafloch. If we take as a starting point the mean temperature of the town of Geneva, which has been determined at 49°·55 F., the elevation of that town being nearly 1,200 feet, we obtain the following approximate results for the mean temperature of the surface at the points in question:--

S. Georges .... 40°·22 Fahr.
S. Livres (Lower) .... 38°·55 "

[214] The English edition of Bischof affords here a proof of the danger of frequent changes from one scale to another. Bischof in the first instance rendered Boussingault into degrees Réaumur, and this was in turn reduced to degrees Fahrenheit; the result being that the authorised English edition of his book gives 2°·25 F. for 127·5 feet, which does not come within 10 feet of Boussingault's statement.

[215] M. Thury calculates a decrease of 1° C. for every 174 mètres between Geneva and S. Bernard, which is less than the decrease given in the text. He arrives at this conclusion by correcting the mean temperature of Geneva from 8°·9 C., the observed mean of eighteen years, to 9°·9 C., in consequence of supposed local causes, which unduly depress the temperature of Geneva. With the mean 8°·9 C. a result nearly in accordance with that of the text is obtained.

Schafloch	.... 33°·88 "
Monthézy	.... 41°·55 "

The law of decrease of temperature enunciated by M. Thury gives a higher mean temperature for the surface of the earth in these places, as in the following table:--

S. Georges	.... 41°·8 Fahr.
S. Livres (Lower)	.... 40°·1 "
Schafloch	.... 35°·6 "
Monthézy	.... 42°·5 "

If any certain information could be obtained of the elevation of the Abbey of Grâce-Dieu, I am sure that a result more surprising than that in the case of the Glacière of Monthézy would appear. The elevation of the floor of the church in the citadel of Besançon is 367·7 mètres, and the plateau on the north side of the town of Baume-les-Dames is 531·9 mètres. I am inclined to think, from the look of the country, that the latter possesses much the same elevation as the valley in which the Abbey lies; and in that case we should have comparatively a very high mean temperature for the surface in the neighbourhood where the glacière occurs.

But if these are the mean temperatures of the surface, the natural temperatures of the caves themselves should be still higher, on account of the allowance to be made for increase of temperature with descent into the interior of the earth. This element will very materially affect our calculations in such a case as the lower part of the ice in the Glacière of the Pré de S. Livres, and the strange suggestive beginning of a new ice-cave 190 feet below the surface, on the Montagne de l'Eau, near Annecy. In any open pit or cave, the ordinary atmospheric influences find such easy access, that the temperature cannot be expected to follow the law observed when perforations of small bore are made in the earth, as in the case of the preliminary boring before commencing to dig a well;[216] but the two glacières mentioned above are so completely protected in their lowest parts, that they may be treated as if they were isolated from external influence of all ordinary kinds; and it may fairly be said that the mean temperature there ought to be considerably higher than at the surface.

It is not very likely that the results of the above calculations are strictly in accordance with what a careful series of observations on the spot might show. The distance between Geneva and the Glacières of S. Georges and S. Livres is sufficiently small to make it probable that the reality is not very far different from the calculated temperature; but the other two caves are comparatively so far off, that the temperature and elevation of Geneva are not very safe data to build upon.

---

[216] Professor Phillips found, in the course of his investigations in the Monk Wearmouth mines, some hundreds of yards below the sea, that when a new face of rock was exposed, its temperature was considerably higher than that of the gallery or shaft in which it lay. In some cases the difference amounted to 9 and 10 degrees. The rock soon cooled down to an agreement with the surrounding temperature.

Appendix

M. Thury's observations during his winter visit to the Glacière of S. Georges are so curious and valuable, that I give the principal results of them here.

It will be remembered that this glacière consists of a roomy cave, 110 feet long and 60 feet high, with two orifices in the higher part of the roof, one of which is kept covered with the trunks of trees to shut out the direct radiation of the sun. A little thought suggested to M. Thury that the cold in the cave in mid-winter would most probably be greater than the external cold of the day, and less than that of the night; so that there should be a time in the later evening when a column of colder and heavier air would begin, to descend through the hole in the roof. To test the correctness of this supposition, he took up his abode in the cavern for the evening of the 10th January, 1858, with a lighted candle. The flame burned steadily for some time; but at 7.16 P.M. it began to flicker, and soon inclined downwards through an angle of about 45°; and when M. Thury placed himself under the principal opening, the flame was forced into an almost horizontal position. At 8 P.M. the current of air had all but disappeared. This violent and temporary disturbance of equilibrium was a matter of much surprise to M. Thury; for he had naturally expected a quiet current downwards, continuing through the greater part of the night.

At 7.16 P.M. the external temperature was 23·9° F., and the temperature of the atmosphere in the cave at the same time was 30°·88 F.;[217] so that there is no wonder the current of air should be strong. It is very difficult to say, however, why it did not commence much earlier, considering that the external air must have been heavier than that in the cave long before 7 o'clock. M. Thury refers to the mirage as a somewhat similar instance, that phenomenon being explained by the supposition that atmospheric layers of different temperatures lie one above another in clearly-defined strata. He suggests, also, that as the heavier air tends to pass down into the cave, the less cold air already in the cave tends to pass out; and the narrow entrance confining the struggle between the opposing tendencies to a very small area, the weaker initial current is able for a time to hold its own against the intruder. On this supposition, it is easy to see that when the rupture does occur it will be violent.

The next day, M. Thury arrived at the glacière at 9.50 A.M. He had determined, in the summer, that the temperature of the cave was invariable, at any rate through the 3-1/2 hours of his visit (from 7.30 to 11 A.M.); but his winter experience was very different. The following are the results of his observations.

In the cave:--

January 9,	at 7.16 P.M.[218]	...	30°·884 Fahr.
"	" 7.20 "	...	29°·75 "
"	" 7.27 "	...	27°·5 "
"	" 7.50 "	...	26°·834 "
January 10,	at 10.12 A.M.	...	23°·684 "
"	" 10.30 "	...	23°·9 "
"	" 11.20 "	...	24°·022 "
"	" 12.14 "	...	24°·134 "
"	" 1.30 "	...	24°·35 "

[217] This was given by a thermometer only placed in the cave at 7 P.M., and by construction not very sensible.

[218] The moment when the disturbance of the atmosphere commenced.

"	" 2.30 "	... 24°·584 "	
"	" 3.14 "	... 24°·8 "	
"	" 4.0 "	... 25°·142 "	

Supposing the weather to have been much the same on the 9th and 10th of January, as M. Thury's account seems to say, there is something very strange in the great difference between the temperatures registered at 4 P.M. on the one day, and at 7.16 P.M. on the other.

The external temperatures at the mouth of the cave were as follows:--

January 10, at	10.53 A.M.	... 25°·934 Fahr.
"	" 11.14 "	... 26°·384 "
"	" 11.45 "	... 28°·04 "
"	" 12.32 P.M.	... 27°·944 "
"	" 1.12	... 30°·644 "
"	" 3.3	... 26°·834 "
"	" 3.56	... 25°·7 "
"	" 4.26	... 25°·25 "

The minimum temperature of the external air during the night of January 10-11 was 18°·392 F., and that of the glacière 19°·76 F.[219] During the preceding night, the minimum in the cave was 22°·442 F., which may throw some light upon the difference between the temperatures at 7.16 P.M. on the 9th, and at 4 P.M. on the 10th.

M. Thury bored a hole, of about 10 inches in depth, in the flooring of ice, and placed a thermometer in it, at 12.25 P.M., closing it up with cotton. At 2.55 P.M., and at 4.7. P.M., the thermometer marked the same temperature, namely, 26°·24 F.

[219] M. Thury gives—4°·62 C. as the minimum in the glacière during the night in question; but on the next page he gives—6°·8 C. (= 19°·76 F.). It is evident, from a comparison with other details of his observations, that the latter is the correct account.

M. Thury's views on glacières in general, based upon the details of the three which he has visited, are much the same as those which I have expressed. He has, however, more belief than I in 'cold currents.'

INDEX

A

B

C

H

I

J

L

M

N

O

P

Q

R

S

T

U

V

W